FM 21-76

DEPARTMENT OF THE ARMY FIELD MANUAL

SURVIVAL

1957 HISTORIC EDITION

HEADQUARTERS, DEPARTMENT OF THE ARMY

Doublebit Press
Eugene, OR

Doublebit Press is an imprint of Eagle Nest Press
www.doublebitpress.com
Eugene, OR, USA

This title, along with other Doublebit Press books are available at a volume discount for clubs or reading groups. Contact Doublebit Press at info@doublebitpress.com for more information.

Military Outdoors Skills Series: Volume 2

Doublebit Press Historic Edition ISBNs
Hardcover: 978-1-64389-016-6
Paperback: 978-1-64389-017-3

First Doublebit Press Historic Edition Printing, 2019

The Military Outdoors Skills Series
Historic Field Manuals and Military Guides
on Outdoors Skills and Travel

Military manuals contain essential knowledge about outdoors life, thriving while in the field, and self-sufficiency. Unfortunately, many great military books, field manuals, and technical guides over the years have become less available and harder to find. These have either been rescinded by the armed forces or are otherwise out of print due to their age. This does not mean that these texts are worthless or "out of date" – in fact, the opposite is true! It is true that the US Military frequently updates its manuals as its protocols have changed based on the times and combat situations that our armed services face. However, the knowledge about the outdoors over the entire history of military publication is timeless!

By publishing the **Military Outdoors Skills Series**, it is our goal at Doublebit Press to do what we can to preserve and share historic military works, such as army field manuals (the FM series), technical manuals (the TM series), and other military books that hold timeless knowledge about outdoors life, navigation, and survival. Through remastered reprint editions of military handbooks and field manuals, we can preserve the time-tested skills and institutional knowledge that was learned through hard lessons and training by the U.S. Military and our expert soldiers.

Soldiers were the original campers, bushcrafters, hikers, backpackers, and survivalists. Because of this, military field manuals about outdoors life contain essential knowledge about thriving in the wilds. This book is an important contribution to outdoors literature and has important historical and collector value toward preserving the American outdoors tradition. The knowledge it holds is an invaluable reference for practicing skills related to thriving in the outdoors. Its chapters thoroughly discuss some of the essential building blocks of outdoors knowledge that are fundamental but may have been forgotten as equipment gets fancier and technology gets smarter. In short, this book was chosen for Historic Edition printing because much of the basic skills and knowledge it contains could be forgotten or put to the wayside in trade for more modern conveniences and methods.

Although the editors at Doublebit Press are thrilled to have comfortable experiences in the woods and love our high-tech and light-weight equipment, we are also realizing that the basic skills taught by the old experts are more essential than ever as our culture becomes more and more hooked on digital stuff. We don't want to risk forgetting the important steps, skills, or building blocks involved with thriving in the outdoors. This Historic Edition reprint represents a chosen group of seminal military handbooks and field manuals that are essential contributions to the American outdoors tradition. In the most basic sense, these books are the collection of experiences by the great experts of outdoors life: our countless expert soldiers who learned to thrive in the backwoods, deserts, extreme cold environments, and jungles of the world.

With technology playing a major role in everyday life, sometimes we need to take a step back in time to find those basic building blocks used for gaining mastery – the things that we have luckily not completely lost and has been recorded in books over the last two centuries. These skills aren't forgotten, they've just been shelved. *It's time to unshelve them once again and reclaim the lost knowledge of self-sufficiency.*

Based on this commitment to preserving our outdoors heritage, we have taken great pride in publishing this book as a complete original work. We hope it is worthy of both study and collection by outdoors folk in the modern era of outdoors and traditional skills life.

Unlike many other photocopy reproductions of classic books that are common on the market, this Historic Edition does not simply place poor photography of old texts on our pages and use error-prone optical scanning or computer-generated text. We want our work to speak for itself, and reflect the quality demanded by our customers who spend their hard-earned money. With this in mind, each Historic Edition book that has been chosen for publication is carefully remastered from original print books, *with the Doublebit Historic Edition printed and laid out in the exact way that it was presented at its original publication.* We provide a beautiful, memorable experience that is as true to the original text as best as possible, but with the aid of modern technology to make as beautiful a reading experience as possible for books that are typically over a century old. Military historians and outdoors enthusiasts alike are sure to appreciate the care to preserve this work!

Because of its age and because it is presented in its original form, the book may contain misspellings, inking errors, and other print blemishes that were common for the age. However, these are exactly the things that we feel give the book its character, which we preserved in this Historic Edition. During digitization, we ensured that each illustration in the text was clean and sharp with the least amount of loss from being copied and digitized as possible. Full-page plate illustrations are presented as they were found, often including the extra blank page that was often behind a plate. For the covers, we use the original cover design to give the book its original feel. We are sure you'll appreciate the fine touches and attention to detail that your Historic Edition has to offer.

For outdoors and military history enthusiasts who demand the best from their equipment, the Doublebit Press Historic Edition reprint of this military manual was made with you in mind. Both important and minor details have equally both been accounted for by our publishing staff, down to the cover, font, layout, and images. It is the goal of Doublebit Historic Edition series to preserve outdoors heritage, but also be cherished as collectible pieces, worthy of collection in any outdoorsperson's library and that can be passed to future generations.

So, without blabbering on further, we hope you enjoy your Doublebit Historic Edition. May your trails be clear and your experiences be memorable!

- The Doublebit Press Editors

FM 21-76

FIELD MANUAL, ⎫   HEADQUARTERS,
⎬   DEPARTMENT OF THE ARMY
No. 21-76, ⎭   WASHINGTON 25, D. C., *25 October 1957*

SURVIVAL

CHAPTER 1

INTRODUCTION

Section I. GENERAL

1. Purpose and Scope

a. Modern combat increases the likelihood of your becoming isolated and having to find water, food, and shelter for many days—even weeks—while making it back to friendly forces. Small units fighting in widely dispersed formations or on special missions forward of friendly lines are more likely to be cut off than ever before. Large units traveling great distances by air and sea make survival in remote and desolate areas a real possibility for you. The ability to evade the enemy and to escape if captured, both basic requirements of the soldier's Code of Conduct, demands every survival skill you can master. The chances of being exposed to such an emergency are always present, especially when traveling, so survival techniques should be a part of your basic soldiering skills.

b. This manual has been written to help you acquire these skills. It tells how to travel, find water and food, shelter yourself from the weather, and care for yourself if you become sick or injured. This information is first treated generally and then applied specifically to such special areas as the Arctic, the desert, the jungle, and on the ocean.

c. Individual skills such as map reading, using a compass or other directional guides, scouting and patrolling, camouflage, first aid, sanitation, personal hygiene, and night vision provide a good foundation on which to build further survival skills. You should have a good general knowledge of them already, so they are repeated in this manual only as they apply to survival specifically.

d. You can remain alive anywhere in the world when you keep your wits. This is a major lesson in survival. Remember that nature and the elements are neither your friend nor your enemy —they are actually disinterested. Instead, it is your determination to live and your ability to make nature work for you that are the deciding factors.

2. Field Skills

A knowledge of field skills, including woodcraft, firemaking, food and water sources, shelter devices, and navigational techniques is necessary for survival. A basic knowledge of woodcraft, for example, prevents wasting valuable time fishing with a hook when a spear or net made from materials at hand would do a better job. Your survival chances increase as your knowledge of field skills increases; as you improve your ability to improvise; and as you learn to apply the principles contained in this manual to your immediate situation.

Section II. INDIVIDUAL AND GROUP SURVIVAL

3. The Will To Survive

a. General. The experiences of hundreds of servicemen isolated during World War II and Korean combat prove that survival is largely a matter of mental outlook, with the will to survive the deciding factor. Whether with a group or alone, you will experience emotional problems resulting from fear, despair, loneliness, and boredom. In addition to these mental hazards, injury and pain, fatigue, hunger, or thirst, tax your will to live. If you are not prepared mentally to overcome all obstacles and accept the worst, the chances of coming out alive are greatly reduced.

b. If you Are Alone. The shock of finding yourself isolated behind enemy lines, in a desolate area or in enemy hands can be reduced or even avoided if you remember the keyword S-U-R-V-I-V-A-L (fig. 1).

    (1) S—size up the situation by considering yourself, the country, and the enemy.

        (a) *Yourself.* Hope for the best but be prepared for the worst. Recall survival training and expect it to work. After all, you have been through this before and the only difference is that this is the real thing. In this way you will increase your chances for success by being confident that you can survive. Get to a safe, comfortable place as quickly as possible. Once there look things over, think, and form a plan. Your fear will lessen; your confidence increase. Be calm. Take it easy until you know where you are and where you are going.

        (b) *The Country.* Part of your fear may come from being

Size up the situation

Undue haste makes waste

Remember where you are

Vanquish fear and panic

Improvise

Value living

Act like the natives

Learn basic skills

Figure 1. Factors for survival.

in strange country; therefore, try to determine where you are by landmarks, compass directions, or by recalling intelligence information passed on to you by your leaders.

(c) *The Enemy.* Put yourself in the enemy's shoes. What

would you do? Watch the enemy's habits and routines. Base your plan on your observation. Remember, you know where the enemy is but he does not know where you are.

(2) *U*—undue haste makes waste.

    (*a*) Don't be too eager to move. It will make you careless and impatient. You begin to take unnecessary risks and you might end up like these men—

        1. "All that was on my mind was to get away, so I just rushed headlong without any plan. I tried to travel at night, but I just injured myself further by bumping into trees and fences. Instead of laying low and trying to evade the enemy, I fired at them with my carbine and was caught the second time."

        2. "I became very impatient. I had planned to wait until night but could not. I left the ditch about noon and walked until I was captured."

    (*b*) Don't lose your temper. It may cause you to stop thinking. When something irritating happens, stop. Take a deep breath and relax; start over.

    (*c*) Face the facts—danger does exist. To try to convince yourself otherwise only adds to the danger. Don't be like the soldier who was captured by a child because he thought, "Capture is the last thing I have to worry about. This is merely a game. It really is not happening to me."

(3) *R*—remember where you are. You may give yourself away because you are used to acting in a certain way. Doing "what comes naturally" may be the tipoff that you don't belong there. One soldier, captured because he whistled a song, reported, "Everything had been going well on the train. Suddenly an ugly little woman started whistling 'Tipperary'. Immediately, I unconsciously began to whistle with her. It gave me away." If he had been one of the "enemy" the chances are he would not have known the song.

(4) *V*—vanquish fear and panic.

    (*a*) To feel fear is normal and necessary. It is nature's way of giving you that extra shot of energy just when you need it. Learn to recognize fear for what it is and control it. Look carefully at a situation and determine if your fear is justified. When you investigate you will usually find many of your fears unreal.

    (*b*) When you are injured and in pain, it is difficult to control fear. Pain sometimes turns fear into panic and causes a person to act without thinking. One pilot, downed during World War II, might have saved himself had he been able to stop and think when his parachute caught in a tree and he was suspended head down, his foot tangled in the webbing. Unfortunately, the pilot's head touched an ant hill and biting ants immediately swarmed over him. In desperation he pulled his gun and fired five rounds into the webbing holding his foot. When he did not succeed in breaking the harness by shooting at it, he placed the last shot in his head.

(c) Panic can also be caused by loneliness. It can lead to hopelessness, thoughts of suicide, and carelessness— even capture or surrender. Recognizing these signs help you overcome panic.

(d) Planning your escape will help keep your mind busy. Find things to do and watch. One soldier, not knowing what to do, decided to kill all of the bugs. There were a lot of spiders, the big ones that do not hurt a human, so he killed the flies and gave them to the spiders to eat. He found something to do. Prayer, reading the Bible, or other religious observance will help calm you. But miracles work best for those who prepare carefully and do all they can to save themselves.

(5) *I*—improvise.

(a) You can always do something to improve the situation. Figure out what you need; take stock of what you have; then improvise.

(b) Learn to put up with new and unpleasant conditions. Keeping your mind on SURVIVAL will help. Don't be afraid to try strange foods. One survivor reported that some men would almost starve before eating strange food. He said they tried a soup made from lamb's head, with lamb's eyes floating around in it. When a new prisoner came in, he would try to find a seat next to him so he could eat the food the prisoner refused.

(6) *V*—value living.

(a) Hope and a real plan for escape reduce your fear and make your chances of survival better. Just beginning to plan his escape to friendly forces made this soldier feel better: "I went outside one time and saw a powerful search light from a distance. I realized this was friendly forces. Immediately I transferred all my thoughts from my personal miseries to escape plans and began to feel better."

(b) Conserve your health and strength. Illness or injury will greatly reduce your chance of survival and escape.

(c) Hunger, cold, and fatigue lower your efficiency and stamina; make you careless; and increase the possibility of capture. Knowing this will make you especially careful because you will realize that your low spirits are the result of your physical condition and not the danger.

(d) Remember your goal—*getting out alive.* Concentrating on the time after you "get out alive" will help you value living now.

(7) *A*—act like the natives. "At the railroad station, there were German guards," one escapee related. "I had an urgent need to urinate. The only rest room was an exposed one in front of the station. I felt too embarrassed to relieve myself in front of all passersby. I walked throughout the entire town occasionally stopping and inquiring if a rest room were available." This man was detected and captured because he failed to accept the customs of the natives. When you are in this situation accept and adopt native behavior. In this way, you avoid attracting attention to yourself.

(8) *L*—learn basic skills. The best life insurance is to make sure that you learn the techniques and procedures for survival so thoroughly that they become automatic. Then the chances are that you will do the right thing, even in panic. Work on the training you are offered because it may mean your life. Be inquisitive and search for additional survival knowledge.

c. *The Group.*

(1) Just as you must make your reactions to survival situations automatic, so must the group that you might be leading. Groups such as squads or platoons that work together and have leaders that fulfill their responsibilities, have the best chance for survival.

(2) If you and your group consider the following factors while evading capture, you should return to friendly forces successfully.

(a) Organize group survival activities.

1. Group survival depends largely upon the organization of its manpower. Organized action where group members know what to do and when to do it, both under ordinary circumstances and during a crisis, prevents panic. One technique for achieving

organized action is to keep your group well-informed. Another is to devise a plan and then stick to it.

2. Assigning each man a task that most closely fits his personal qualifications is an additional way of organizing your group. If one man feels he can fish better than cook, assign him the job of providing fish. Always try to determine and use special skills within the group.

3. Panic, confusion, and disorganization are minimized by good leadership. It is your responsibility as the senior member of the group to assume command and establish a chain of command that includes all members of the group. Make certain that each man knows his position in the chain of command and is familiar with the duties of every other man, especially yours. Under no circumstances should leadership of the group be left up to chance acceptance by some member after a situation arises.

(b) Lead your men. Group survival is a .test of effective leadership. Maintain your leadership prestige by using it wisely; be the leader, set the example. Supervise constantly to prevent serious arguments, troublemakers from attracting undue attention, those who may "crack up" from disrupting the group, and to prevent carelessness caused by fatigue, hunger, and cold. Know yourself and your men and be responsible for each individual's welfare.

(c) Develop a feeling of mutual dependence within the group by stressing that each man depends on the other men for survival. Emphasize that wounded or injured men will not be left behind—that it is each member's responsibility to see that the group returns intact. This attitude fosters high morale and unity. Each member receives support and strength from the other.

(d) No matter what the situation, the leader must make the decisions. Because he needs intelligence upon which to base his decision, he should ask for information and advice from other members of the group—much as a general officer uses his staff. Above all else, the leader must at all times avoid the appearance of indecision.

(e) Situations arise that must be acted upon immediately. The ability to think on your feet usually determines survival success. Consider the facts and make decisions rapidly.

4. Avoiding Detection

Survival when you are isolated in enemy territory depends as much on your ability to avoid detection and capture as it does to find enough food, water, and shelter. You must know—

a. How to conceal yourself when the enemy is near and to move without silhouetting yourself against the skyline; and how to keep from being spotted from enemy aircraft.

b. How far noises carry in fog, falling snow, heavy foliage, or over rock faces.

c. How smells from cooking food, tobacco, wood smoke, body odors, and body wastes can reveal your location.

d. The dangers of sudden, rapid movement.

e. How to observe the enemy without being observed.

f. How to camouflage yourself, your camp, and equipment; and the dangers of using too much camouflage.

g. How to select routes for movement which avoid exposed areas; how to move quietly without leaving obvious tracks; and how to determine travel time for yourself or for a group.

h. How to signal using your voice, hands and arms, pebbles, and pieces of wood.

5. Suppose You're Captured?

a. What happens if you become a prisoner of war? After all, it is possible. Isolation, fear, injury—all work in favor of the enemy to increase the chances of capture, in spite of a determined effort on your part to evade. The surrender of your arms, however, doesn't mean that you forfeit your responsibilities as an American soldier. The Armed Forces Code of Conduct directs that you begin planning your escape the minute you are taken prisoner.

b. Escape is tough; making it stick is even tougher. It demands courage and cunning and much planning—of seeking ways out, routes to follow, and the location of friends. Above all, it demands physical stamina—stamina that you must acquire under the worst conditions imaginable. Experience has proved that "model" camps, where rations are regular and treatment considerate, are the exception. But no matter what extremes your life as a POW assumes, your aim should be the same—to keep yourself physically able and sufficiently equipped for the breakout.

6. A Plan for Survival

Since the conditions in various POW camps differ, it is impossible to provide a specific survival plan for each situation. What you need is a guide to help you plan to make the best of what you have. Here is one such plan that you can remember by the word S-A-T—*Save, Add to, Take care of.*

a. Save. What can you *save* in a POW camp? Everything—clothing, pieces of metal, cloth, paper, string—anything. A piece of twine may mean success or failure when it comes time to break out. Hide these items under the floor or in a hole in the ground. If they are discovered, they may appear harmless and little or nothing will be done to punish you.

(1) Wear as few clothes as possible. *Save* your shoes, underwear, shirts, jacket, and any other items of clothing that will protect you from the elements when you begin your trip back.

(2) *Save* any nonperishable foods that you receive from the Red Cross or your captors. Candy, for example, comes in handy as a quick source of energy when traveling. If no other candy source is available, *save* each issue of sugar given you by the enemy. When you get enough, boil it down into hard candy. *Save* it until you build up your supply. Canned foods that you might receive are ideal for storing. However, if the enemy punctures the cans to prevent your saving it, you may still preserve this food by resealing the cans with wax or some other field expedient. It may be feasible for you to save this food by recooking it and changing its form. Other foods to hoard against the day of your escape include suet and cooked meat, nuts, and bread.

(3) *Save* pieces of metal no matter how insignificant they may seem. Nails and pins can serve as buttons or fasteners. Old tin cans are excellent for improvised knives, cups, or food containers. If you are fortunate enough to have a razor blade, guard it. Use it for shaving only. Devise ways of sharpening it—rub it on glass or stone or some other hard surface. A clean shave is a good morale booster.

(4) *Save* your strength but keep active. A walk around the compound or a few mild calisthenics keep your muscles toned. Sleep as much as you can. You won't get much rest on your way back.

b. Add To.

(1) Use your ingenuity. Select those items that you can't get along without and supplement them; for example, your rations. There is more to eat, in and around your compound, than you think. When you are allowed to roam around the camp grounds, look for natural foods

native to the area. See chapters 4 and 6 for a discussion of edible foods. If possible, *add* these roots, grasses, leaves, barks, and insects to your escape cache. They will keep you alive when the going gets tough.

(2) Supplement your clothing so the more durable garments are in good repair when you escape. A block of wood and a piece of cloth make good moccasins and save your boots. Rags can substitute for gloves; straw can be woven into hats. Don't forget to salvage clothing from the dead.

c. *Take Care Of.* Probably the most important part of any plan for survival is the "take-care-of" phase. Maintain what you have. There won't be any reissue when your shoes wear out or you lose your jacket. Also, it's easier to maintain good health than to regain it once it's lost.

(1) Put some of your clothing into your escape cache. Watch the rest for early signs of wear and repair it with improvised material, if necessary. A needle made from a thorn, nail, or splinter and threaded with unraveled cloth, can mend a torn pair of trousers. Wood canvas, or cardboard bound to the soles of your shoes will save them from wear. Even paper will suffice as a reinforcing insole if your shoes do wear through.

(2) Good physical health is essential to survival under any circumstances. It is especially important in a POW camp where living conditions are crowded and food and shelter inadequate. This means that you must use every device possible to keep yourself well.

(a) Soap and water is a basic preventive medicine; so keep clean. If water is scarce, collect rainwater, use dew, or simply rub yourself daily with a cloth or your bare hands. Pay attention to areas on your body that are susceptible to rash and fungus infection—between your toes, your crotch and scalp.

(b) The cleanliness rule also applies to your clothing. Use soap and water when you can spare it. Hang your clothes in the sun to air if soap and water are not available. Examine the seams of your clothing and hairy portions of your body frequently for lice and their eggs. Disease infected lice can kill. A possible way to get laundry service or even a bath is to tell your guard that you are infested with lice, whether or not your complaint is true. The prison authorities, fearing that lice on prisoners may cause an outbreak of louse-borne disease among the civilian population, might provide this service.

(c) In the event you become ill, report your condition to the camp authorities. The chance that you will receive aid is worth the try.

(d) For additional tips on maintaining your health, see paragraphs 7 through 9.

Section III. HEALTH AND FIRST AID

7. You Are the Doctor

a. Keeping well is especially important when you are on your own. Your physical condition has much to do with your will to survive and your success in returning safely.

b. Chances are you will not have trained medical personnel to help you maintain your health. You must rely on your initiative and your knowledge of first aid to prevent disease and to treat any injuries that you might receive.

8. Aids to Maintaining Health

Protecting yourself against disease and injury involves making habits of a lot of simple rules which we call personal hygiene. The immunizations you have had will continue to give you good protection against a few of the more serious diseases to which you may be exposed: smallpox, typhoid fever, tetanus (lockjaw), typhus, diphtheria, and cholera. They will *not* protect against the much more common diseases like diarrhea, dysentery, colds, and malaria. Your only means of preventing these is by keeping physically fit and by keeping disease germs out of your body. Applying the following rules will go a long way toward keeping you on your feet.

a. *Keep Clean.*

(1) Body cleanliness is the first defense against disease germs. A daily shower with hot water and soap is ideal. If this is impossible, keep the hands as clean as possible and sponge the face, armpits, crotch, and feet at least once a day.

(2) Keep your clothing, especially your underclothing and socks, as clean and dry as possible. If laundering is impossible shake out your clothing and expose to sun and air daily.

(3) If you have a toothbrush, use it regularly. Soap or table salt and soda make good substitutes for toothpaste, and a small green twig, chewed to a pulpy consistency at one end, will serve as a toothbrush. After eating

rinse your mouth if purified water is available.

b. *Guard Against Intestinal Sickness.*

(1) Common diarrhea, food poisoning, and other intestinal diseases are the commonest if not the most serious of the diseases you will have to guard against. They are caused by putting filth or poisons into the mouth and stomach. To guard against these diseases—

(a) Keep the body, particularly the hands, clean. Keep your fingers out of your mouth. Avoid handling food with the hands.

(b) Purify your drinking water by use of purification tablets or by boiling for four minutes. Avoid beverages from native sources.

(c) Avoid eating raw foods, especially those grown on or in the ground. Wash and peel fruits.

(d) Avoid holding food for long periods following preparation.

(e) Sterilize your eating utensils by heat.

(f) Keep flies and other vermin off your food and drink. Keep your camp clean.

(g) Adopt strict measures for disposing of human wastes. Apparently healthy persons can carry deadly germs.

(2) If you develop vomiting or diarrhea, rest and stop eating solid foods until the symptoms ease up. Take fluids, particularly water, in small amounts at frequent intervals. As soon as can be tolerated, resume eating semi-solid foods. Normal salt intake should be maintained.

c. *Guard Against Heat Injury.* In hot climates develop a tan by gradual exposure to the sun. Avoid strenuous exertion in the hot sun; you may develop fatal heat stroke. The lesser illnesses caused by heat can be prevented by consuming enough water and salt to replace the sweat. Salt tablets or table salt should be taken in the proportion of 2 tablets or ¼ teaspoonful to a quart of water. Treatment of heat casualties consists of cooling the body and restoring water and salt.

d. *Guard Against Cold Injury.*

(1) When exposed to severe cold conserve your body heat by every means possible. Take particular care of the feet, hands, and exposed parts. Keep your socks dry and use any available material including rags and paper to improvise protective covering.

(2) Frostbite is a constant danger to anyone exposed to temperatures below freezing. Treatment of frostbite consists of getting the patient into a warm place (normal room temperature) as soon as possible; rapidly rewarm-

ing frozen parts of the body by immersion in warm water (90° F. — 104° F.); by placing a warm hand on the frozen part; or by exposure to warm air. *Do not massage* or apply snow or ice to the affected area. See paragraphs 58*b* and *c*.

e. Guard Against Insects and Insect-Borne Diseases. Common insects such as flies, mosquitoes, lice, ticks, and mites carry many of our most serious diseases such as typhoid fever, dysenteries, malaria, brain fever, and yellow fever. Every possible means should be used to avoid the contamination of food by flies and the bites of mosquitoes and other insects. Lacking screening, bednets, insecticides, and repellents, this will be very difficult since improvisations are hard to come by. Some of the things you can do are—

(1) Protect food and beverages from flies and other vermin.
(2) Avoid close contact with natives.
(3) Cover the body to reduce exposure to mosquitoes, especially after dark.
(4) Take a suppressive drug to prevent malaria when such is available.
(5) Keep free of lice.
(6) Remove ticks promptly.

f. Guard Against Contact Diseases. Many diseases such as venereal diseases, dysenteries, tuberculosis, measles, mumps, and common respiratory and skin diseases result from close association with cases or carriers of these diseases. Avoid intimate contacts with the native who is likely to be a source of these diseases.

g. Take Care of Your Feet.

(1) Do not wear dirty or sweaty socks. If you don't have a clean pair, wash out those that you have on. If you have an extra pair to wear, put the washed pair inside your shirt next to your body. They will dry in a short time. If possible, wear woolen socks; they absorb perspiration.
(2) Blisters are dangerous because they may cause fatal infections. If your shoes fit well and you dry them after crossing wet ground, if you change your socks frequently, and if you exercise your feet, you should not have much trouble with blisters. Should you develop a blister, however, pierce through the thick skin at its base with a sterilized needle or knife blade (the point of the needle or knife may be quickly sterilized by holding it in a flame for a few seconds); press and drain the blister. Then apply a clean bandage to prevent the dead skin from being rubbed off before it is healed.

9. Survival First Aid

a. Illness and injury are always potential survival partners. Situations may arise when you will have to treat yourself or your companions. You are not likely to have either adequate medical supplies or trained personnel to administer treatment. You will have to improvise equipment to perform even simple first aid.

    (1) You can make bandages and dressings fairly sterile by boiling or steaming them in a covered container or charring them.

    (2) Try to keep all improvised equipment as sterile as possible. Extreme heat is your best method of doing this.

b. Medical treatment other than the simplest first aid can be dangerous. If you don't know the procedure for treatment, make the person comfortable rather than chance injuring him further with improper treatment.

c. Injuries that you are most likely to encounter while on your own include cuts and bruises, fractures, sprains, concussions, and burns. In all cases of severe injury, keep the person lying down; if he is unconscious, keep him lying on his side or on his belly, with his head turned to one side to prevent choking. Handle the injured man carefully, especially if he has a fracture or back injury. Give as much treatment for shock as possible.

d. Follow these procedures—

    (1) *Shock.*

        (a) This condition is characterized by paleness, trembling, sweating, and thirst, and can accompany any injury. The more severe the wound, the more likely it is that shock will develop.

        (b) If he is unconscious lay the patient flat on his back. Raise his feet unless he has a head injury or breathing difficulty. Keep him comfortably warm but avoid overheating him. If the patient is conscious give him warm drinks.

        (c) If you are alone and become seriously injured, lie down in a depression in the ground, behind a tree, or any place sheltered from the wind. If you can, lie with your head lower than your feet to increase the flow of blood to your head. Keep yourself as warm as possible and rest for at least 24 hours.

    (2) *Bleeding.* Stop bleeding as soon as possible by using the following methods:

        (a) If you have a first aid packet, place the sterile dressing

directly on the wound and press it with your hand, or bandage the wound firmly.

(b) If the bleeding is from an arm or leg, and if bleeding continues, elevate the injured area and continue the pressure.

Caution: if you suspect a broken bone, do not elevate the leg or arm.

(c) If bleeding continues in spite of pressure dressing and elevation of injured arm or leg, then apply finger pressure on appropriate pressure point shown in figure 2.

(d) Apply a tourniquet only if you are unable to control the bleeding by applying pressure and by elevation. In cases where severe, uncontrolled bleeding necessitates the use of the tourniquet, it should be applied between the wound and the heart. In cases where there is a traumatic amputation (loss or an arm, leg, hand, foot, etc.), place the tourniquet near the end of the stump. In all other cases where a tourniquet is required to control bleeding, place it above the elbow or knee. Once you apply a tourniquet, do not loosen or release it regardless of how long it has been on. A tourniquet, once applied, should be removed only by medical personnel who are trained and equipped to control the bleeding by other means and to restore the lost blood volume. But remember—the tourniquet should be regarded only as a last resort in control of bleeding! It should never be used if the bleeding can be adequately controlled by pressure and elevation.

(3) *Fractures.*

(a) Handle persons with fractures carefully to avoid causing additional injury.

(b) If the fracture is accompanied by a wound, tear or cut away the man's clothing and treat the wound before splinting it.

(c) Splint the patient before moving him. Improvise splints from branches, a tight roll of clothing, or pieces of equipment. Pad the splint and place it so that it supports the joints above and below the fracture. Immobilize a fractured leg by tying it to the unbroken leg, if other materials are unavailable.

(4) *Sprains.*

(a) Bandage and rest the sprained limb.

(b) Apply cold applications for the first 24 hours after injury; then apply heat.

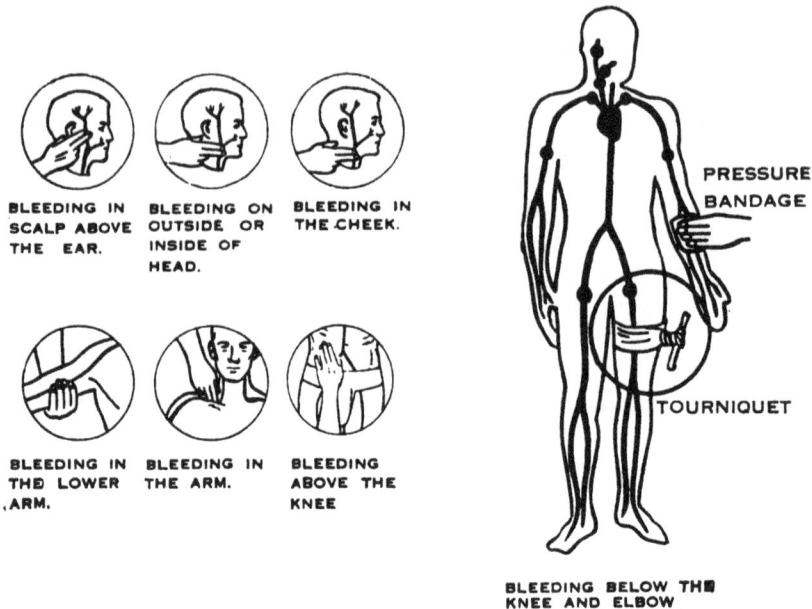

BLEEDING IN SCALP ABOVE THE EAR.

BLEEDING ON OUTSIDE OR INSIDE OF HEAD.

BLEEDING IN THE CHEEK.

BLEEDING IN THE LOWER ARM.

BLEEDING IN THE ARM.

BLEEDING ABOVE THE KNEE

PRESSURE BANDAGE

TOURNIQUET

BLEEDING BELOW THE KNEE AND ELBOW

Figure 2. Pressure points.

(c) If it is necessary to use the sprained limb, splint the injured area as much as possible. Sprained limbs can be used to the limit that pain will allow.

(5) *Concussion.*

(a) Skull fractures or other head injuries should be suspected if there is unconsciousness, thin watery blood or blood-tinged water escaping from the nose or ears, convulsion, or unequal or unresponsive pupils of the eyes. The aforementioned signs are frequently accompanied by headache and vomiting.

(b) Keep the patient warm and dry and handle him gently.

(6) *Burns.*

(a) Sunburn will probably be the chief type of burn injury you will encounter. Protect the person from further exposure, and cover the sunburned area with ointment or a substitute made by boiling the bark of an oak, hemlock, or chestnut tree.

(b) Do not touch a sunburned area. Apply the ointment and cover the burn with a dressing. Do not remove the bandage except as an emergency or due to excessive soiling. Give the patient large amounts of fluid. Keep the burned area at rest.

e. For treatment of cold injuries, see paragraph 58.

CHAPTER 2

ORIENTATION AND TRAVELING

Section I. Navigation

10. Where Are You?

One of the first survival problems you must solve is determining where you are and in what direction you must go to reach friendly forces.

a. Isolation on Land Close to the Frontlines. Should you be cut off from your unit as a result of enemy action, try to remember the location of friendly forces. Then, travel in that direction, using the sun and stars as directional guides.

b. Isolation in a Desolate Area or Deep in Enemy Territory.

(1) If you are aboard a plane which is forced down over land, you can keep from getting lost by knowing the direction of flight and something about the country over which you were flying. Prior to or during the flight, try to study available maps and photographs of the area, the direction of flow of large rivers, the direction in which mountains or ridges run, and the location of outstanding landmarks with relation to friendly forces.

(2) If you are abandoning a ship at sea or an aircraft over a large body of water and time permits, find out—

(a) Your course and direction to the nearest land.

(b) The latitude and longitude.

(c) Prevailing wind direction.

(d) Direction of flow of ocean currents.

(3) Should you be an escapee from a PW enclosure deep in enemy territory and find yourself lost, seek a hideout; sit down, relax, and think over the situation. Try to recall landmarks you saw on your trip to the enemy rear.

11. Finding Yourself Using a Map

You are fortunate if you have a map with which you can determine your location, a possible route to safety, and natural obstacles that you will encounter. However, before you use your map be sure it is oriented. You can do this by inspection or with a compass.

    a. Inspection.

        (1) Climb the nearest hill or tree. Look at the surrounding country, then at your map.

        (2) Turn the map until the roads, rivers, hills, or woods around you look as if they are in the same place on your map as they are on the ground. When your map lines look much the same as the land lines, your map is turned in the right direction—the north side of the map is toward north, the east side to the east.

    b. With a Compass.

        (1) Lay the map flat on the ground or other smooth, level surface.

        (2) Place your compass on the map and turn the map until the north-south grid lines are parallel to the compass needle and north coincides with compass north.

        (3) Turn the map again until the needle on the compass indicates the amount of magnetic declination for the area. The magnetic declination diagram usually appears in the marginal information.

12. Guiding by Sun and Stars

    a. Sun.

        (1) Early morning or late evening is not the only time that you can determine direction from the sun's position. During winter in the Northern Hemisphere, the path of the sun is south of the zenith, a position directly overhead. During summer its path is almost directly overhead. At noon, during the winter months, the sun is due south. Shadows at this time of day are thrown north. If you are isolated in the Southern Hemisphere, the exact opposite is true.

        (2) Pay attention to where the sun hits you when you face the direction that you want to travel. Check its position to your travel direction frequently as it will change with the time of day and season of the year.

          (a) An ordinary watch can be used to determine the approximate true north. In the North Temperate Zone only, the hour hand is pointed toward the sun. A north-south line can be found midway between the hour hand and 12 o'clock. This applies to standard time; on daylight saving time, the north-south line is found midway between the hour hand and 1 o'clock. If there is any doubt as to which end of the line is

NORTH TEMPERATE ZONE

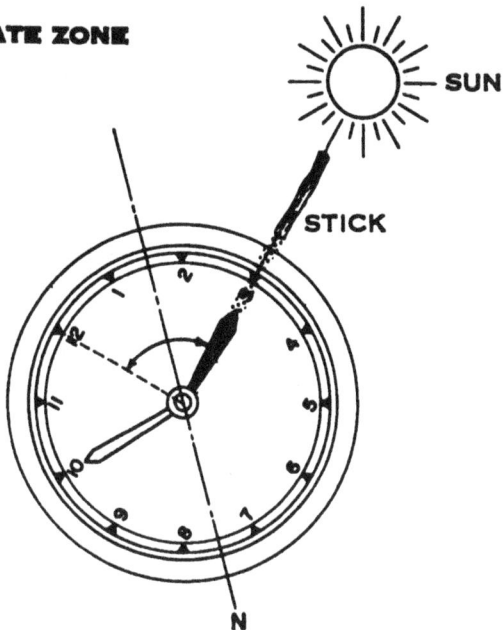

SUN

STICK

N

SOUTH TEMPERATE ZONE

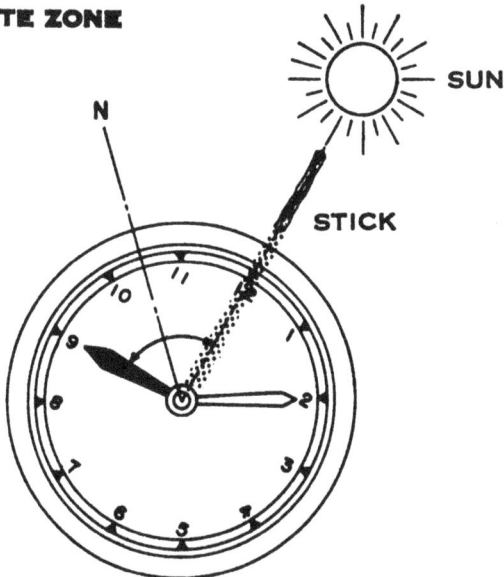

SUN

N

STICK

Figure 3.   Using a watch to find north.

north, remember that the sun is in the eastern part of the sky before noon and in the western part in the afternoon. The watch may also be used to determine direction in the South Temperate Zone. However, it is used a bit differently. Twelve o'clock is pointed toward the sun, and halfway between 12 o'clock and the hour hand will be a north-south line. If on daylight saving time, the north-south line lies midway between the hour hand and 1 o'clock. The temperate zones extend from latitude 23½° to 66½° in both hemispheres (fig. 3).

(b) On cloudy days, place a stick at the center of the watch and hold it so that the shadow of the stick falls along the hour hand. One-half the distance between the shadow and 12 o'clock is north (fig. 4).

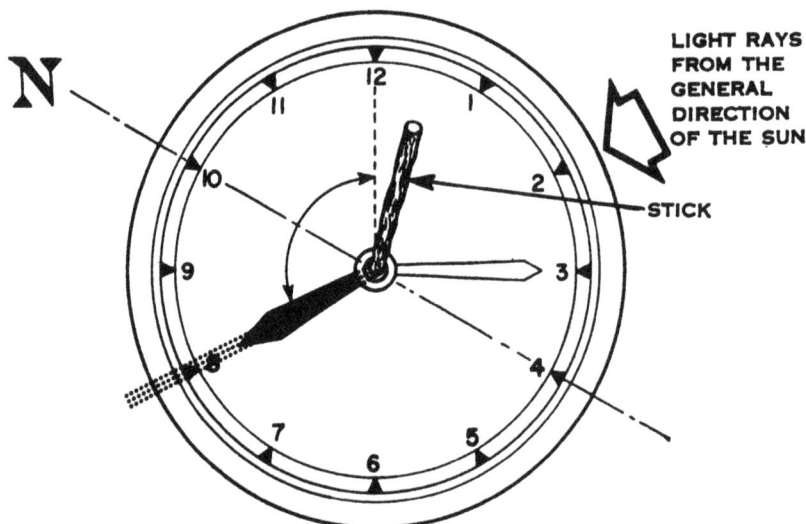

Figure 4. Finding north on cloudy days.

b. *Stars.*

(1) In the Northern Hemisphere a line from any observer to the North Star (Polaris) is never more than 1° away from true north. To locate the North Star—

(a) Find the Big Dipper (fig. 5).

(b) Extend an imaginary line through the two pointer stars which form the side of the cup farthest from the handle. About five times the distance between these two stars in the direction from which you would pour from the dipper, is the North Star.

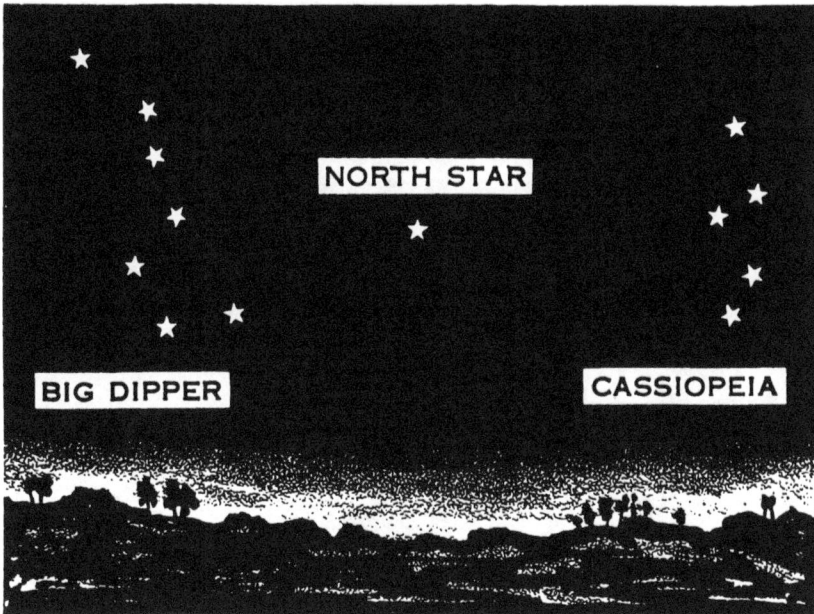

Figure 5.   The Big Dipper.

(2) In the Southern Hemisphere you can find south by locating the Southern Cross (fig. 6). Compare this group of stars to a kite. If you can figure the length of the kite from tip to tail and extend an imaginary line from the tip of the tail four and one-half times the length of the kite, you can determine the approximate direction of south.

13. Keeping a Course

When clouds obscure the sun or stars, you need other methods to help you maintain your direction.

a. In strange country, study outstanding terrain features as you travel, and concentrate on keeping your course. Climb to a high point and look at the general pattern of the land, character of the vegetation, the drainage patterns, and the trend of mountains and ridges. Choose a prominent landmark that you can see while you travel. As you near this landmark, line up another one.

b. If you are traveling in a dense forest, you probably won't be able to spot distant landmarks. You can hold a course by lining up two trees forward of your position in your direction of travel. As soon as you pass the first one, line up another beyond the second. You might find it helpful to look back occasionally to

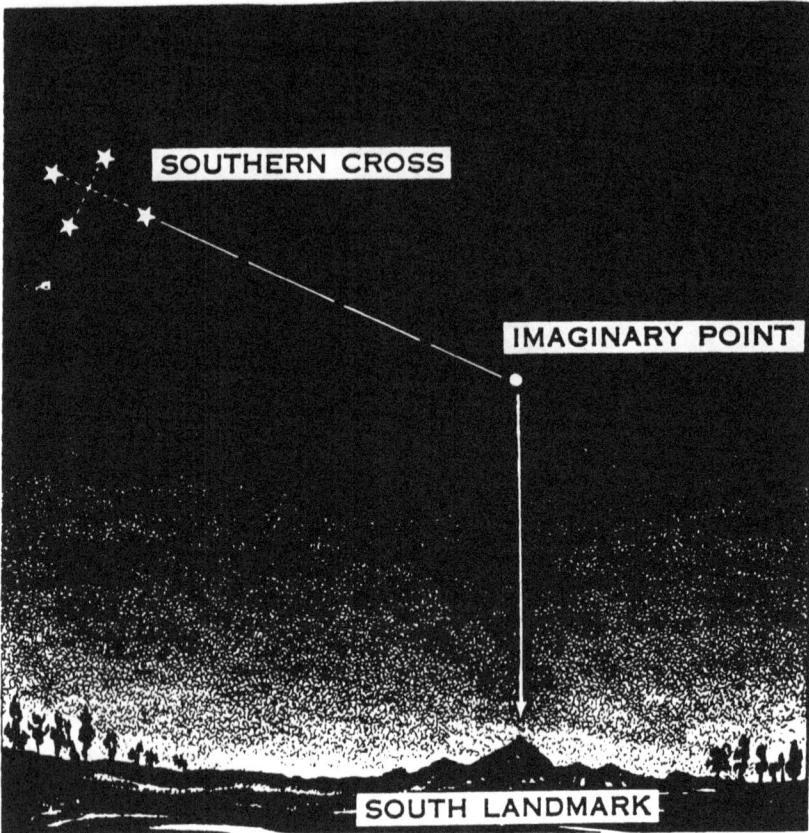

Figure 6. Southern Cross.

check the relative positions of landmarks or ground slope and contour.

c. You can usually use streams, ridges, and trees as guides in open country and as a means of retracing your route. On overcast days, in areas where the vegetation is dense, or whenever the country appears the same, mark your route with bent bushes, rocks, or notches cut in tree trunks. Make bushmarks by cutting vegetation or bending it so that the under and lighter side of the leaves is facing upward. These signs are especially conspicuous in dense vegetation but should be used with discretion because of the risk of discovery involved in plainly marking your route.

d. Even if you have a map, don't guide too confidently on manmade features or landmarks that are likely to change. The only safe landmarks are natural features such as rivers and hills. In the jungle, for example, when a village site marked on a map is investigated, it will often be an overgrown clearing.

Similarly, one rainy season can change the course of a small stream or close an unused trail with dense shrub.

e. Guide on trails that lead in your general direction, and when you come to a fork, guide on the path that appears most traveled: If you guide on the wrong trail and find yourself lost, stop and try to remember the last time that you were sure of where you were. Mark your location and start "back tracking." Sooner or later you will discover a recognizable feature with which you can pinpoint your position.

f. Travel at night is safe in the desert or open country. but not advisable in strange, wooded country. However, if you do travel at night, use a shielded light only when necessary to find your way over rough, dangerous spots, or to read a map or compass. Your eyes adjust to the darkness; a light blinds you to all but a small area that is illuminated. You can keep a fairly accurate course for short distances in open country by picking a bright star near the horizon (guide star) in your line of travel and lining it up with trees and other skyline landmarks ahead. Be sure you check your direction frequently with the North Star or Southern Cross and change your guide star accordingly.

g. You might have to detour frequently in rough country. Use the following methods to get back on your course:

(1) In short detours, estimate the distance and average angle of departure. On your return, gauge the angle and distance so as to strike your line again. For greater accuracy, count paces and use a compass (fig. 7)[1].

(2) Select a prominent landmark ahead and behind your line of travel. On return from your detour, walk until you are again "lined up" on the two landmarks; then follow your original course (fig. 8).

(3) An easy way to compensate is by paces and right angles, although it requires more walking (fig. 9).

Section II. SELECTING YOUR ROUTE ON THE GROUND

14. Study the Terrain

The route that you select to travel depends upon the situation in which you find yourself, the weather conditions, and the nature of the terrain. Whether you select a ridge, stream, valley, coastline, dense forest, or mountain range to follow, be sure it is the safest, rather than the easiest way. Experience has proved that the most difficult route is frequently the safest.

---

[1] Aviation Training, Office of the Chief of Naval Operations, United States Navy, "How To Survive on Land and Sea," Copyright 1943, 1951 by The United States Naval Institute.

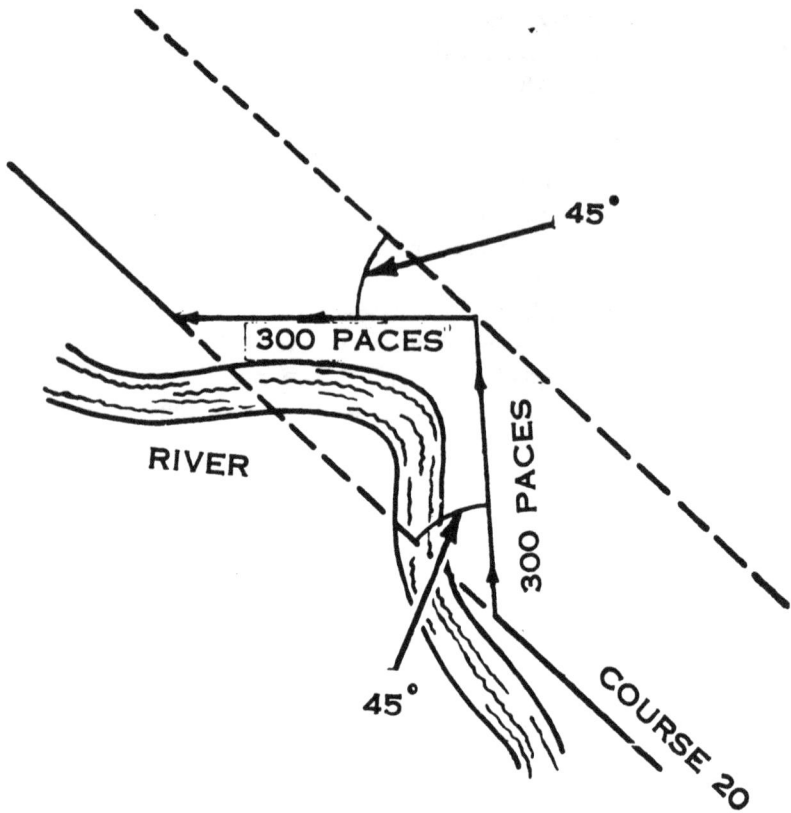

Figure 7. Estimating distance and average angle of departure.[2]

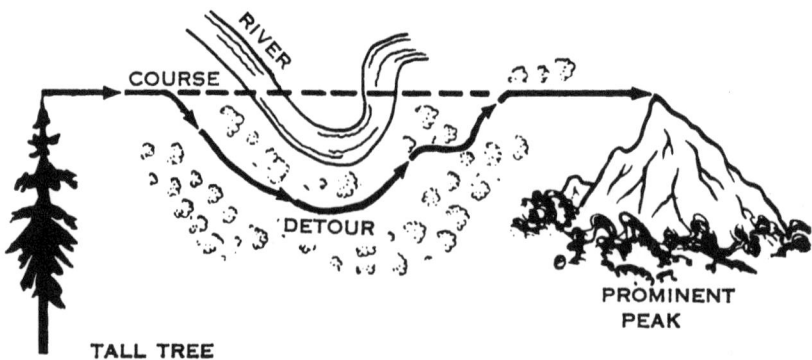

Figure 8. Using a prominent landmark.[3]

[2]Ibid.
[3]Ibid.

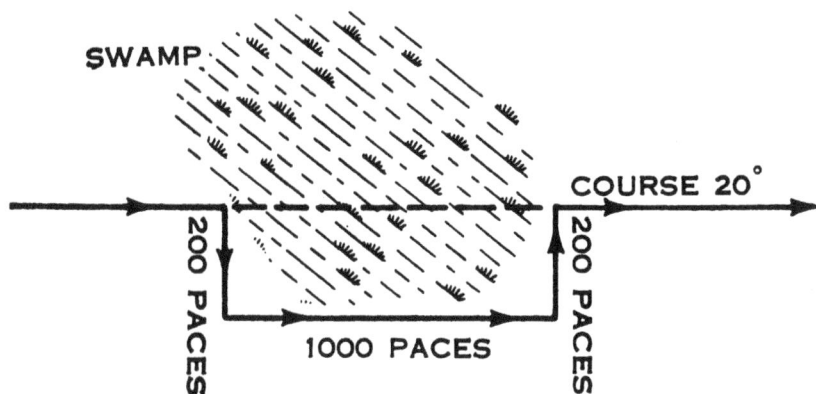

Figure 9. Compensating by paces and right angles.[4]

15. Following a Ridge

A route along a ridge line is usually easier to follow than one through a valley. Game trails are frequently on top of ridges and you can use them to guide your travel. Also you find less vegetation, frequent high points for observing landmarks, and few streams and swamps to ford.

16. Following Streams

Using a stream as a route is of particular advantage in strange country because it provides a fairly definite course and might lead to populated areas; it is a potential food and water source and a vehicle for travel by boat or raft. However, be prepared to ford, detour, or cut your way through the thick vegetation lining the stream. If you are following a stream in mountainous country, look for falls, cliffs, and tributaries as check points. In flat country streams usually meander, are bordered by swamps, and are thick with undergrowth. Travel on them provides little opportunity to observe landmarks.

17. Following Coast or Shorelines

If you decide to follow a coastline you can figure on a long, roundabout route. But it will be a good starting line, an excellent baseline from which to get your bearings, and a probable source of food.

18. Through Dense Vegetation

a. With practice you can move through thick undergrowth and jungle fairly silently by cautiously parting the vegetation to make your way.

---
[4] Ibid.

b. Avoid scratches, bruises, and loss of direction and confidence by developing "jungle eye." Disregard the pattern of trees and bushes directly in front of you. Focus your eyes beyond your immediate front, and rather than looking *AT* the jungle, look *THROUGH* it. Stoop occasionally and look along the jungle floor.

c. Keep alert. Move slowly and steadily in dense forests or jungles, but stop periodically to listen and take your bearings. You cover more territory, and birds and animals do not reveal your position by their cries.

d. Use a bayonet or machete to cut your way through dense vegetation, but do not cut unnecessarily. The noise caused by chopping carries a long distance in the woods. You can reduce this noise by stroking upward when cutting creepers and bush.

e. Many jungle and forest animals follow well established game trails. These trails wind and crisscross but frequently lead to water or clearings. If you use these trails do not follow them blindly. Make sure they lead in your direction of travel by checking your bearings frequently.

f. When you climb a tree to observe or get food, be sure you test each limb before putting your weight on it, and always have a good handhold on something sturdy. Climb close to the trunk because limbs are strongest at this point (fig. 10).

19. Mountains

a. Travel in mountains or other rugged country can be dangerous and confusing unless you know a few tricks. What looks like a single ridgeline from a distance might be a series of ridges and valleys. In extremely high mountains, a snowfield or glacier that appears to be continuous and easy to travel over might cover a sheer drop of hundreds of feet. In jungle mountains, trees growing in valleys formed by streams reach great heights, and their tops are at about the same level as trees growing on the valley slopes and hilltops where the water is scarce; this forms a tree line that from a distance appears level and continuous.

b. Follow valleys or ridges in mountainous terrain. If your route leads to a hidden gorge with walls almost straight up and down, search for a bypass.

c. To save time and energy during mountain walking, keep the weight of your body directly over your feet by placing the soles of your shoes flat on the ground. If you take small steps and move slowly but steadily this is not difficult.

    (1) When you ascend hard ground—

        (a) Lock your knees briefly at the end of each step in order to rest your leg muscles.

Figure 10. Climbing close to the trunk.

(b) Traverse steep slopes in zigzag direction—not straight up.

(c) Turn at the end of each traverse by stepping off in the new direction with the up-hill foot. This prevents crossing your feet and possibly losing your balance.

  (2) When you descend hard ground—

    (*a*) Come straight down without traversing.

    (*b*) Keep your back straight and your knees bent so that they take up the shock of each step.

    (*c*) Keep your weight directly over your feet by placing the full sole on the ground at each step.

  d. You may have to go up or down a steep slope and cliff. Before you start, pick your route carefully, making sure it has places for hand or foot holds from top to bottom. Try out every hold before you put all of your weight on it, and distribute your weight. Keep the following hints in mind:

  (1) If possible, don't climb on loose rock.

  (2) Move continually, using your legs to lift your weight and your hands to keep your balance.

  (3) Be sure you can go in either direction without danger at any time.

  (4) In climbing down, face out from the slope as long as possible. This is the best position from which to choose your routes and holds.

  (5) Rappel (*e* below) whenever you are descending steep slopes, if an easier route for descent is not available.

  e. When you are traveling in mountainous country, make an effort to acquire a rope and an ice axe, or you will find the job of descending steep slopes difficult or even impossible. If necessary, improvise one from parachute shroud lines.

  (1) Rappelling (fig. 11).

    (*a*) Loop the rope around a tree or rock, allowing the ends to hang evenly.

    (*b*) Put both ropes between your legs.

    (*c*) Wrap them around your right (left) thigh as shown in figure 11.

    (*d*) Pass the ropes across your chest, over your left (right) shoulder, and down across your back, grasping them with your right (left) hand.

    (*e*) Take hold of the ropes in front of you with your free hand.

    (*f*) Relax your grip periodically and slide down slowly in a rhythmic manner. You must keep yourself perpendicular to the slope or cliff to prevent your feet from sliding off the rock.

      1. Keep your feet spread and braced against the cliff.

      2. Slow or stop yourself by tightening your grip on the ropes and bringing your grasping hand across your chest.

Figure 11. Rappelling.

3. When you reach bottom, pull one strand of the doubled rope to retrieve it.

(2) *Alternate method of rappelling.* If you have one or more companions and desire a safer method of descending a cliff, have one man act as a belayer to give additional security to the man rappelling. Tie the rope around the waist of your companion by means of a bowline or a bowline on a bight (fig. 12). Get into a secure sitting position with the feet braced. The rope is passed low around the waist and is paid out as the man moves. You must remember that the man moving should have freedom of action and you should hold him tight only when he asks for it.

f. As you travel down a grade, be on the lookout for slopes of loose, relatively fine rock. These slopes can aid your movement. Turn slightly sideways, keeping your body relaxed, and descend diagonally in long jumps or steps. If the slope consists of large rocks, move slowly and carefully to prevent a boulder from rolling under your weight. Always step squarely on the top of a rock to prevent it from throwing you off balance.

20. Snowfields and Glacier Travel

a. The quickest way to descend a steep snowfield is to slide down on your feet, using an ice axe or a stout stick about five feet long as a brace and to dig into the snow to stop your fall if you stumble. You can use the stick to probe for deadly crevasses that may lay beneath innocent looking snow.

b. If you are traveling on a glacier, you must expect to find crevasses (cracks in the ice) generally at right angles to the direction of glacier flow. Usually it is possible to travel around them since they seldom extend completely across the glacier. If snow is present, the greatest caution must be exercised, and party members should be tied into a rope so as to secure each other. Heavily crevassed areas should be avoided whenever possible. Unless you are experienced in glacier travel or unless you can find no other route you should avoid glaciers—they are dangerous for the untrained (fig. 13). See FM 70-10.

c. Travel up or across a steep slope covered with snow is easier if you kick steps into it as you move diagonally across it; but, be on the alert for avalanches, especially during a spring thaw or after a fresh snow. If you must move where there is danger of avalanches, stay out of the valley away from the base of the slope or if you must cross the slope, cross as high as you can. If you must climb the slope, climb straight up. If caught in an avalanche, use swimming motions to stay on top.

Figure 12. Two-loop bowline.

d. Another hazard to travel in mountainous snowfields is projections that are formed by snow blowing from the windward side of a ridge. The projections, or cornices, will not support your weight. You can usually spot them from the leeward side, but from the windward you may see only a gently rounded snow-covered ridge. Follow the ridge on the windward side well below the cornice line (fig. 14).

21. Crossing Water

a. *General.*

   (1) Unless you are traveling in the desert, there is a good possibility that you will have to ford a stream or river. The water obstacle may range from a small, ankle-deep brook that flows down a side valley to a rushing, snow- or ice-fed river. If you know how to cross such an ob-

Figure 13. Using a stick to traverse a snowfield.

stacle, you can use the roughest of waters to your advantage. However, before you enter the water check the temperature. If it is extremely cold and if a shallow fording place cannot be found, it is not advisable to try

Figure 14. Snow cornice.

to cross by fording. The cold water may easily cause a severe shock, which can temporarily paralyze you. In this case, try to make an improvised bridge by felling a tree over the stream or build a simple raft.

(2) Before you attempt to ford, move to high ground and examine the river for—

　(*a*) Level stretches where it breaks into a number of channels.

　(*b*) Obstacles on the other side that might hinder your travel. Pick a spot on the opposite bank where travel will be easier and safer.

　(*c*) A ledge of rocks that crosses the river, indicating the presence of rapids or canyons.

　(*d*) Any heavy timber growths. These indicate where the channel is deepest.

(3) When you select your fording site, keep the following points in mind:

　(*a*) When possible, choose a course leading across the current at about a 45° angle downstream.

　(*b*) Never try to ford a stream directly above or close to a deep or rapid waterfall or deep channel.

　(*c*) Always ford where you would be carried to a shallow bank or sandbar should you lose your footing.

　(*d*) Avoid rocky places, since a fall can cause serious injury; however, an occasional rock that breaks the current may help you.

b. Methods of Crossing.

(1) *Wading.* Before you enter the water remove your shoes and socks unless you need them to protect your feet from being cut by sharp rocks or sticks. Use a stout pole for support. It makes your footing more secure. Also use the pole to test the stream for potholes.

(2) *Swimming.*

(a) Use the breast, back, or side strokes. They are noiseless, less exhausting than other techniques, and will allow you to carry small bundles of clothing and equipment as you swim. If possible, remove your clothing and equipment and float it across the river. Wade out until the water is chest deep before you begin swimming. If the water is too deep to wade, jump in feet first with your body straight; keep your legs together and your hands at your sides. In deep, swift water, swim diagonally across the stream with the current.

(b) If you are unable to swim, you can ford a river by using certain swimming aids. These include—

1. *Clothing.* Take off your trousers in the water; knot each leg and button the fly. Grasp the waist band on one side and swing the trousers over your head from back to front so that the waist opening is brought hard down on the surface of the water. Air is trapped in each leg (fig. 16). If you are not worried about noise, hold your trousers in front of you and jump into the water (fig. 15). Either of these methods provides a serviceable pair of water wings.

2. *Empty tins, gas cans, and boxes.* Lash these together as shown in figures 17, 18, and 19. Use them only when crossing slow moving water.

3. *Logs or planks.* Before you decide to use a wooden floating aid, test its ability to float. This is especially important in the tropics because most tropical trees sink, particularly the palm, even when the wood is dead.

(3) *Rafts.*

(a) Rafting rivers is one of the oldest forms of travel and often is the safest and quickest method of crossing a water obstacle; however, building a raft under survival conditions is tiring and time consuming unless you have proper equipment and help. With these two

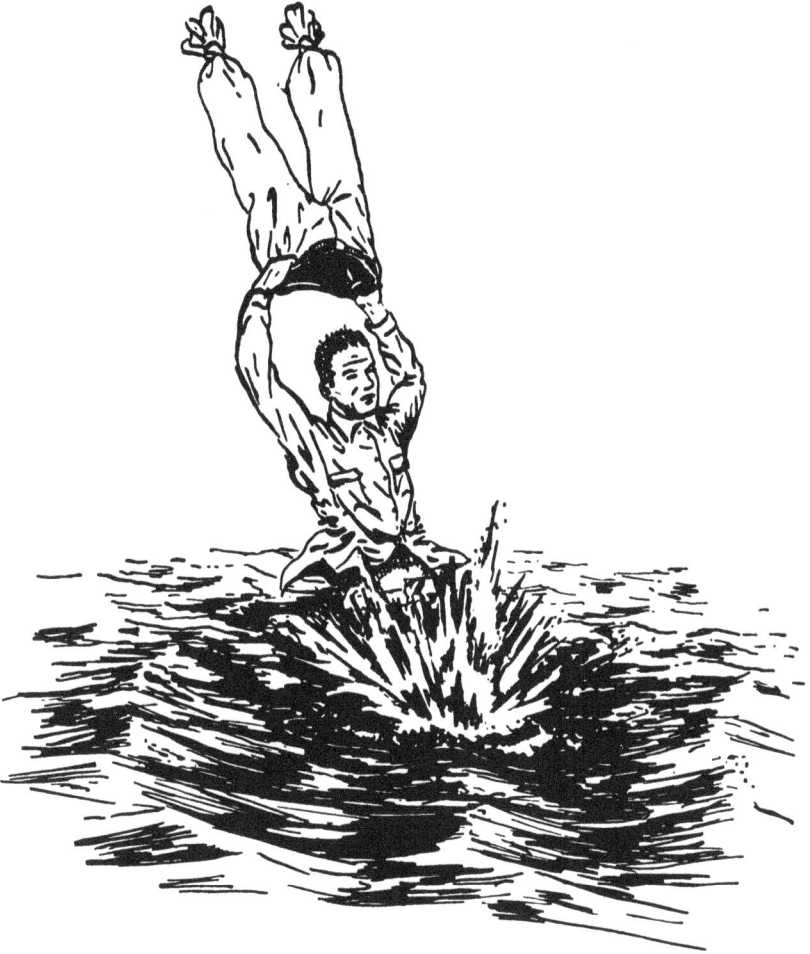

Figure 15. Inflating pants.

requirements you can make rafts from dry standing trees, bamboo, or brush.

(b) Spruce trees that are found in polar and subpolar regions make the best rafts. You can construct a raft without spikes or rope. All you need is an axe and knife. Considering a suitable raft for three men to be 12 feet long and 6 feet wide—

1. Build the raft on two skid logs placed so they slope downward to the bank. Smooth the logs with an axe so the raft logs lie evenly on them.

2. Cut four offset, inverted notches, one in the top and bottom of both ends of each log (fig. 20). Make the

Figure 16. Pants as water wings.

notches broader at the base than at the outer edge of the log.

3. To bind the raft together, drive through each notch a three-sided, wooden crosspiece about a foot longer than the width of the raft (fig. 20). Connect all the notches on one side of the raft before connecting those on the other.

4. Lash the overhanging ends of the two crosspieces together at each end of the raft to give it additional strength. When the raft enters the water the cross-

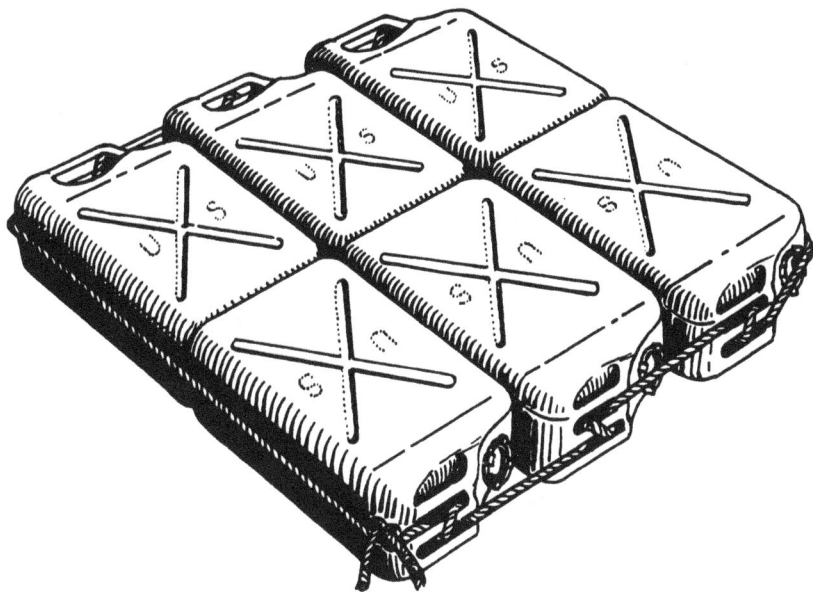

Figure 17. Gas cans.

Figure 18. Tying boxes together.

pieces swell and bind the logs together tightly.

5. If the crosspieces fit too loosely, wedge them with thin pieces of dried wood. These swell when wet, tightening and strengthening the crosspieces.

(c) Bamboo is light, tough, and cuts easily. It makes a serviceable raft.

(d) With a tarpaulin, shelter half, or other waterproof

Figure 19.   Using a crate.

material, you can build an excellent raft using brush as a frame. See FM 5-10.

(e) In Northern Europe, during the winter, rivers may be open in the middle part because of the swift current, and the frozen shores. Cross such a river on an ice block raft which can be cut off from the frozen shore ice, using an ax or even sometimes a pole (if there is a crack in the ice). The size of the raft should be about 2 x 3 yards and the ice should be at least one foot thick. A pole is used to move the ice block raft across the open part of the river (fig. 21).

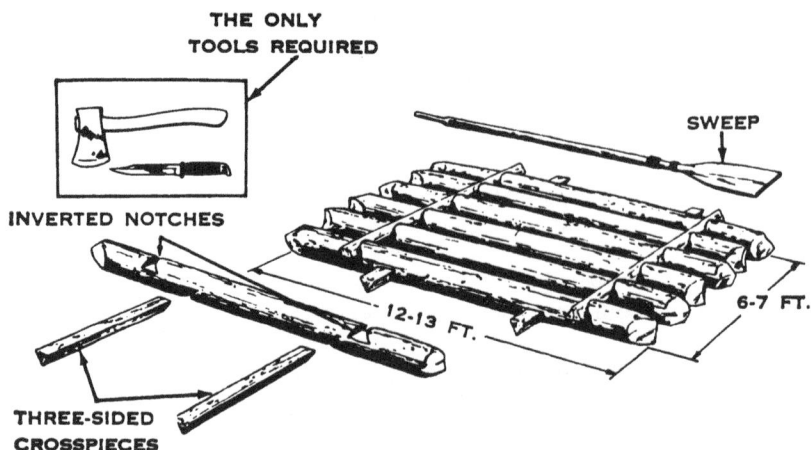

THE ONLY
TOOLS REQUIRED

SWEEP

INVERTED NOTCHES

THREE-SIDED
CROSSPIECES

12-13 FT.

6-7 FT.

Figure 20.   Constructing a log raft.

Figure 21. Ice block raft.

c. *Rapids or Swift Water.*

    (1) Swimming in rapids or swift water is not as great a problem as you think. In shallow rapids, get on your back with your feet pointing downstream; keep your body horizontal and your hands alongside your hips. Flap your hands much like a seal moves his flippers. In deep rapids, swim on your stomach and aim for shore when possible. Watch for currents that converge; you might be sucked under because of the swirls they produce.

    (2) A raft crossing of a deep and swift river may be effected by utilizing a pendulum action at a bend in the river (fig. 22). This method is useful when several men have to cross.

d. *Surf.* You stand a better chance of surviving in surf if you

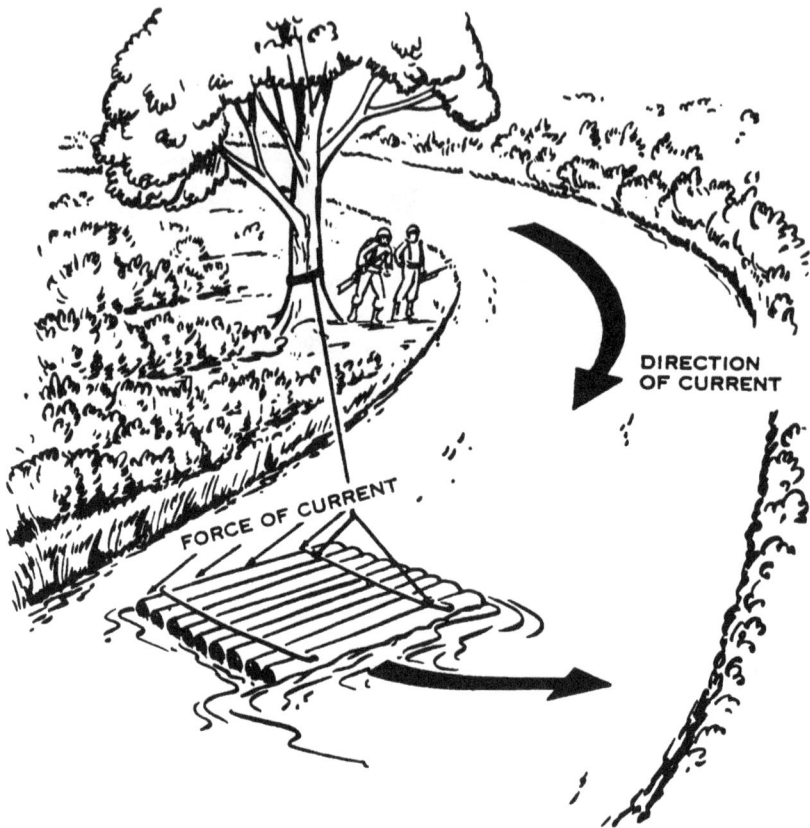

Figure 22. Crossing a river by pendulum action.

know a little about it. Breaking waves become higher and shorter as they move shoreward. The side facing the shore curves and forms the breaker, actually moving the water toward the shore. Large waves break farther away from the shore than do smaller ones.

    (1) In moderate surf, swim forward with the small waves; the crests pick you up. Dive when your ride on the crest ends and just before the wave breaks.

    (2) In heavy surf, swim toward the shore while you are between the waves in a trough. As another wave comes toward you, face it, dive, and swim forward after it passes.

    (3) The backwash of incoming waves can be dangerous if the waves are large. If you are caught in this outbound current don't try to swim against it. Swim with it. If you are carried under, push upward from the bottom,

or swim to the surface and ride toward shore on the next incoming wave.

e. Quicksand, Bogs, Quagmire. These obstacles are found most frequently in tropical or semitropical swamps. Pools of muck are devoid of any visible vegetation and usually will not support even the weight of a rock. If you cannot detour such an obstacle, attempt to bridge it using logs, branches, or foliage. If none are available, cross it by falling face downward with your arms spread. Start swimming or pulling your way through, keeping your body horizontal. Use the same method for crossing quicksand (fig. 23).

Figure 23. Crossing muck or quicksand.

22. Signaling While on Your Way

There is always the chance that you might be rescued, but one man, or a group, is not too easy to spot from the air, especially when visibility is limited. Therefore, you must be prepared to make your whereabouts and needs known to rescuers.

a. Tramp out letters in the snow or use branches to spell your message. If you are on sand, use boulders or seaweed (fig. 24). Select material whose color contrasts with the surface you lay it on. Use the signal code illustrated in figure 25.

b. Produce smoke. Make a large fire and pile enough damp vegetation on it to smother it.

c. Use clothing as flags. Wave your undershirt, shorts, or

Figure 24. Tramping out a signal.

trousers or spread them against a contrasting background.

d. Flash a beam of light from a mirror or other shiny material. Improvise a mirror from a ration tin or belt buckle. Punch a crosshole in the center of the reflector. Reflect sunlight from the mirror to a nearby surface and slowly bring it to eye level, and look through the sighting hole. You will see that a bright spot of light is on the target. Continue sweeping the horizon even though you see no ships or planes. Mirror flashes can be seen for miles even on hazy days (fig. 26).

e. Use a spruce torch for night signaling. Select a tree with dense foliage. Place dry timber in the lower branches to light the tree.

1 REQUIRE DOCTOR, SERIOUS INJURIES

2 REQUIRE MEDICAL SUPPLIES

3 UNABLE TO PROCEED

4 REQUIRE FOOD AND WATER

5 REQUIRE FIREARMS AND AMMUNITION

6 REQUIRE MAP AND COMPASS

7 REQUIRE SIGNAL LAMP WITH BATTERY AND RADIO

8 INDICATE DIRECTION TO PROCEED

9 AM PROCEEDING IN THIS DIRECTION

10 WILL ATTEMPT TAKEOFF

11 AIRCRAFT SERIOUSLY DAMAGED

12 PROBABLY SAFE TO LAND HERE

13 REQUIRE FUEL AND OIL

14 ALL WELL

15 NO

16 YES

17 NOT UNDERSTOOD

18 REQUIRE ENGINEER

NOTE: A SPACE OF 10 FEET BETWEEN
ELEMENTS, WHENEVER POSSIBLE

Figure 25.   Ground-air emergency code.

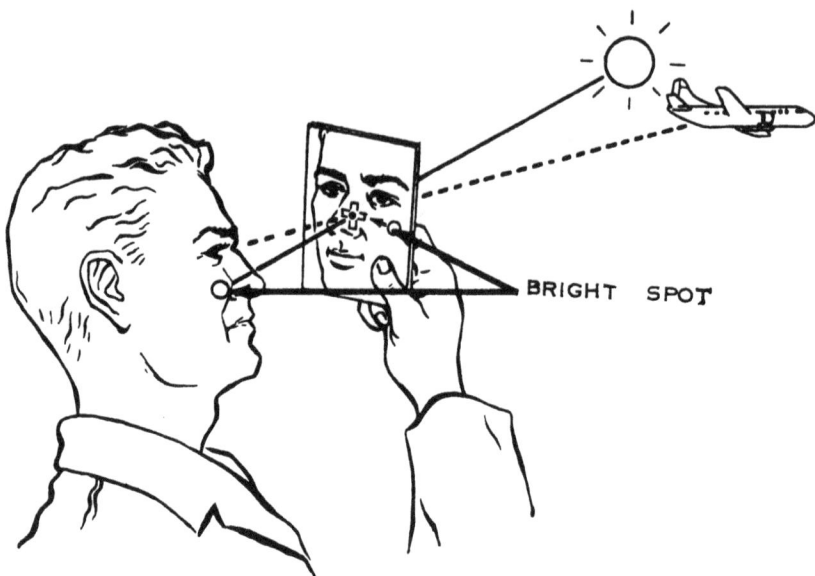

BRIGHT SPOT

Figure 26. Signaling with a mirror.

CHAPTER 3

WATER

Section I. GENERAL CONSIDERATIONS

23. Your Greatest Need

a. The Greek philosopher Miletus was obviously thinking of survival when he declared, "The first of things is water." He couldn't have been more correct. Without water your chances of living are nil, and all the food in the area means nothing. This is especially true in hot climates where you sweat a lot. Even in cold weather your body needs at least two quarts of water each day; any lesser amount reduces your efficiency.

b. Learn to use your water intelligently. Where water is scarce, drink sparingly. When you are extremely thirsty or if you are hot from exercise, sip small amounts of water at a time. But drink your fill. By all means don't drink urine—the waste material in it will make you sick. Purify all water, if at all possible.

24. Drinking Impure Water Is Dangerous

a. No matter how overpowering your thirst may seem, don't drink impure water. You are inviting disaster. One of the worst hazards to survival is waterborne diseases. Impure water teems with disease organisms. Purify all your water either by boiling for at least one minute or by using water purification tablets.

b. Some of the diseases you may contract by drinking impure water include dysentery, cholera, and typhoid.

    (1) *Dysentery.*

        (a) You can be sure you have dysentery if you experience severe and prolonged diarrhea with bloody stools, fever and weakness.

        (b) Eat frequently and try drinking coconut milk, boiled water, or the juice of boiled bark. Eat boiled rice if it's available.

    (2) *Cholera and typhoid.* When you are in an area known for the prevalence of cholera and typhoid fever, be care-

ful even though you may have had innoculations against these diseases.

c. Impure water may also contain flukes and leeches. Drinking these with water can have severe consequences. See chapter 7.

(1) *Flukes.* Blood flukes exist in stagnant, polluted water, especially in tropical areas. If you swallow one while drinking, the fluke will bore into your bloodstream, live as a parasite, and cause painful, often fatal diseases The best protection is to boil your drinking water. Flukes (worm parasites) also may penetrate the unbroken skin while a person is wading or bathing in contaminated water.

(2) *Leeches.*

(a) Small leeches are particularly prevalent in African streams. If you swallow one, it will hook itself to your throat passage or inside your nose. While in this position it will suck blood, create a wound, and move to another area. Each new wound will continue to bleed, opening the door for infection.

(b) Sniff highly concentrated salt water to remove these parasites, or pick them out with improvised tweezers.

25. Muddy, Stagnant, and Polluted Water

a. If you have exhausted all other sources and are still without water, you may drink from a muddy or stagnant pool, even though it may have an odor and be otherwise unpleasant. Be sure to boil this water for at least one minute.

b. To clear muddy water, let it stand for 12 hours, or —

(1) Pass it through about three feet of bamboo that you have filled with sand. Stuff grass in one end to keep in the sand.

(2) Pour it into a cloth that has been filled with sand.

c. Boil polluted water and add charcoal from the fire to get rid of odors. Let the water stand for about 45 minutes before you drink it.

Section II. FINDING WATER

26. What Can You Drink?

a. When you can find no surface water, tap through the earth's water table for ground water—rain or melted snow that has sunk into the ground. Access to this table and its supply of generally pure water depends upon the contour of the land and the character of the soil (fig. 27).

(1) *Rocky soil.*

    (a) Look for springs and seepages. Limestones have more and larger springs than any other type rock. Because limestone is easily dissolved, caverns are readily etched in it by ground water. Look in these caverns for springs.

    (b) Because lava rock is porous, it is a good source for seeping ground water. Look for springs along the walls of valleys that cross the lava flow.

    (c) Be on the lookout for seepage where a dry canyon cuts through a layer of porous sandstone.

    (d) In areas abundant with granite rock, look over the hillsides for green grass. Dig a ditch at the base of the greenest area, and wait for the water to seep in.

Figure 27. Water table.[5]

(2) *Loose soil.*

    (a) Water is usually more abundant and easier to find in loose soil than in rocks. Look for ground water along valley floors or on the slopes bordering the valley because it is in these areas that the water table is most likely to surface. Land above river valleys also yield springs or seepages along their bases, even when the stream is dry.

    (b) If you decide to dig for water, first look for signs that it is present. Dig in the floor of a valley under a steep slope, or dig out a green spot where a spring was during the wet season. In low forests, along the seashore, and in river plains, the water table is close to the surface. Very little digging usually yields a good supply of water.

    (c) Runoff water is found above the water table and includes streams, stagnant pools, and water in bogs. Consider it contaminated and dangerous even if it is away from human habitation. Boil or treat this water with water purification tablets before you drink it.

---

[5]Aviation Training, Office of the Chief of Naval Operations, US Navy, "How To Survive on Land and Sea," Copyright 1943, 1951 by the United States Naval Institute.

27. Along the Seashore

a. You can find water in the dunes above the beach or even in the beach itself. Look in hollows between sand dunes for visible water, and dig if the sand seems moist. On the beach, scoop holes in the sand at low tide about 100 yards above the high tide mark. This water may be brackish, but it is reasonably safe. Run it through a sand filter to reduce the brackish taste.

b. Don't drink sea water. Its salt concentration is so high that body fluids must be drawn to eliminate it. Eventually your kidneys will cease functioning.

28. In Desert or Arid Lands

a. Watch for water indicators when you are isolated in desert or arid regions. Some of the signals include the direction in which certain birds fly, plants, and converging game trails.

    (1) The sand grouse of Asia, crested larks, and Zebra birds visit water holes at least once a day; parrots and pigeons must live within reach of water. Note the direction in which these birds fly and chances are you will find something to drink.

    (2) Cattails, greasewoods, willows, elderberry, rushes, and salt grass grow only where ground water is near the surface. Look for these signs and dig. If you don't have a bayonet or entrenching tool, dig with a flat rock or sharp stick.

b. Desert natives often know of lingering surface pools in low places. They cover them in various ways, so look under likely brush heaps or in sheltered nooks, especially in semiarid and brush country.

c. Places that are visibly damp, where animals have scratched, or where flies hover indicate recent surface water. Dig here for water.

d. Collect dew on clear nights by sponging it up with your handerchief. During a heavy dew you should be able to collect about a pint an hour.

e. For a detailed discussion of finding water in the desert, see chapter 6.

29. On Mountains

Dig in dry stream beds. Water is often present under the gravel. If you are in mountain snowfields, put snow in a container and place it in the sun out of the wind. Improvise your tools from flat rocks or sticks if you have no digging equipment.

30. Water From Plants

If you are unsuccessful in your search for ground or runoff water, or if you do not have time to purify the questionable water, a water-yielding plant may be your best bet. Clear, sweet sap from many plants is easily obtained. This sap is pure and chiefly water. Check the following sources in an emergency:

a. *Plant Tissues.*
   (1) Many plants with fleshy leaves or stems store drinkable water. Try them wherever you find them.
   (2) The barrel cactus of the southwestern United States is a possible source of water (fig. 28). Use it only as a last resort and only if you have the energy to cut through the tough, outer spine-studded rind. Cut off the top of the cactus and smash the pulp within the plant. Catch the liquid in a container. One survivor sliced the pulp into chunks and carried it in his pocket as an emergency water source. A barrel cactus 3½ feet high will yield about a quart of milky juice. *This is the exception to the rule that milky or colored sap-bearing plants should not be eaten.*

b. *Roots of Desert Plants.* Desert plants often have their roots near the surface. The Australian "water tree", desert oak, and bloodwood are some examples. Pry these roots out of the ground and cut them into 24-to 36-inch lengths. Remove the bark and suck out the water.

c. *Vines, Palms, Coconuts.*
   (1) *Vines.* Not all vines yield palatable water, but try any vine you find. Use the following method for tapping a vine. It will work on any species.
      (a) Cut a deep notch in the vine as high up as you can reach.
      (b) Cut the vine off close to the ground and let the water drip into your mouth or a container.
      (c) When the water ceases to drip, cut another section off the top. Repeat this until the supply of fluid is exhausted (fig. 29).
   (2) *Palms.* Buri, Coconut, Sugar, and Nipa palms (figs. 30, 31, 32, and 33) contain a drinkable sugary fluid. To start the fluid of coconut palm flowing, cut off the tip of the flower stalk after bending it downward. If you cut off a thin slice every 12 hours, you can renew the flow and collect up to a quart a day.
   (3) *Coconut.*
      (a) Select green coconuts. You can open them easily

Figure 28. Water from a barrel cactus.

with a knife, and they have more milk than ripe coconuts. Be careful not to drink more than three or four
cups of ripe coconut juice a day. This juice is a
violent laxative.

(*b*) To open a coconut without a knife, drive a stick into
the ground and sharpen the protruding end. Bring
the nut down on the point with enough force to crack
the outer fibrous covering. Smash the hard inner
shell against a tree or rock (fig. 34).

31. Plants That Catch and Hold Water

a. Bamboo stems often have water in the hollow joints. Shake
the stems of old, yellowish bamboo. If you hear a gurgling, cut
a notch at the base of each joint and catch the water in a container
(fig. 35).

b. In the American tropics, the overlapping, thickly growing
leaves of the pineapple-like bromeliads (fig. 36) may hold a considerable amount of rain water. Strain the water through cloth
to eliminate most of the dirt and water insects.

c. Other water-yielding plants include the traveler's tree of
Madagascar (fig. 37); the umbrella tree of western tropical
Africa; and the baobab tree of northern Australia and Africa
(fig. 38).

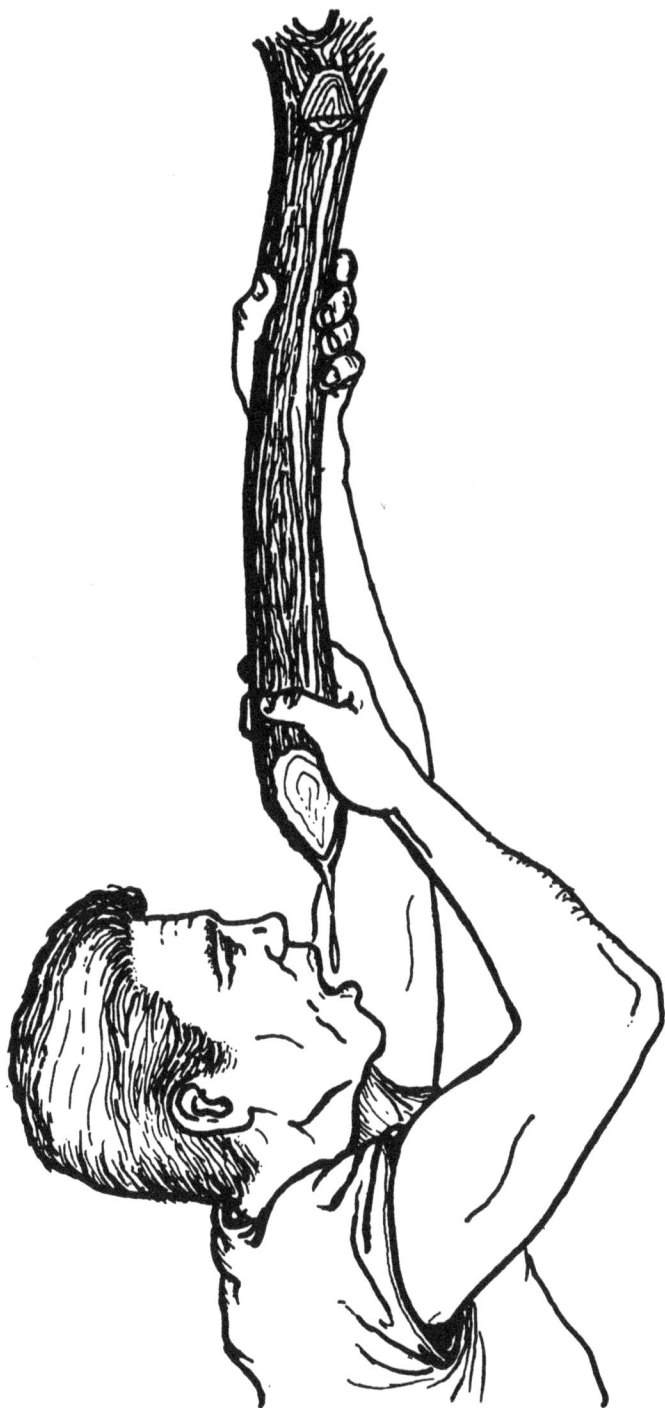

Figure 29. Extracting water from vines.

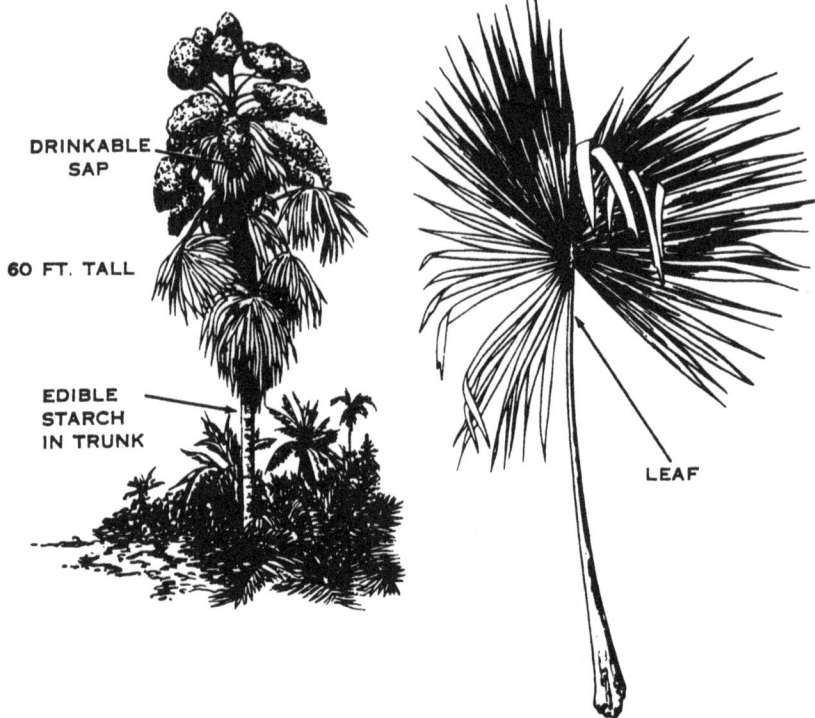

DRINKABLE SAP

60 FT. TALL

EDIBLE STARCH IN TRUNK

LEAF

Figure 30.   Buri palm.

Figure 31. Coconut palm.

FLOWER STALK
(SAP)

15-20 FT. TALL
60 FT. AT MATURITY

CROSS SECTION
OF PALM NUT

EDIBLE
PALM NUT

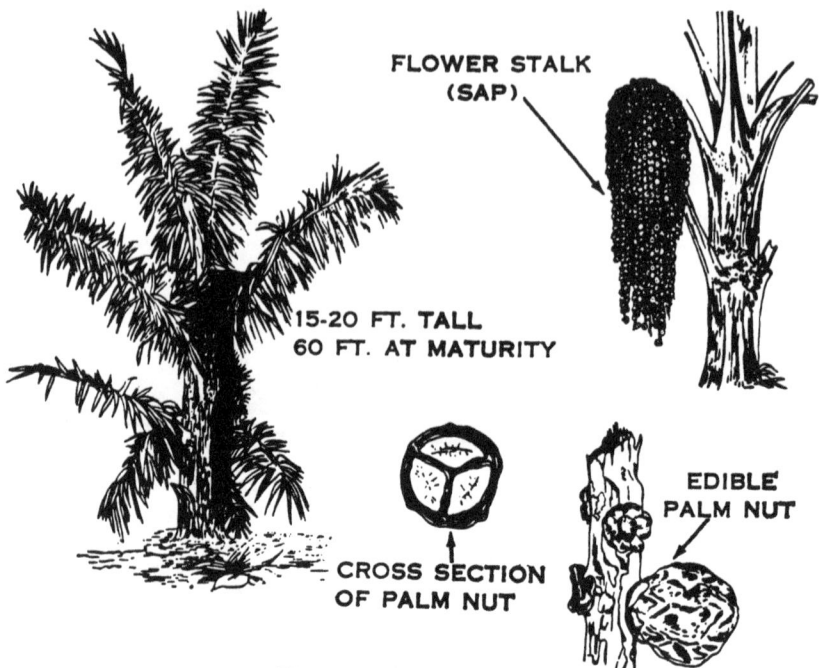

Figure 32. Sugar palm.

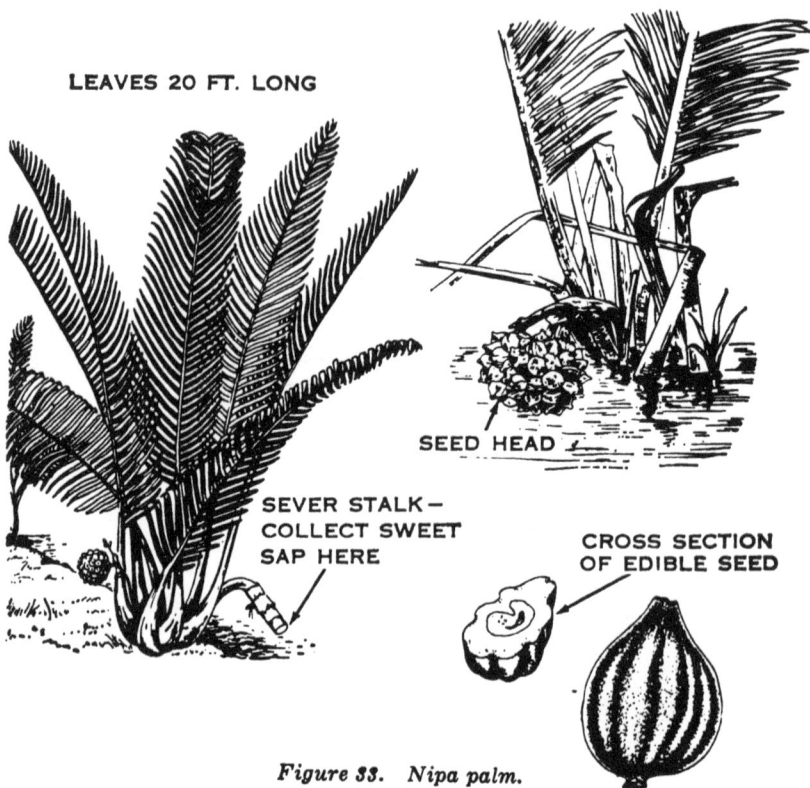

LEAVES 20 FT. LONG

SEED HEAD

SEVER STALK—
COLLECT SWEET
SAP HERE

CROSS SECTION
OF EDIBLE SEED

Figure 33. Nipa palm.

Figure 34. Using a stick to crack a coconut.

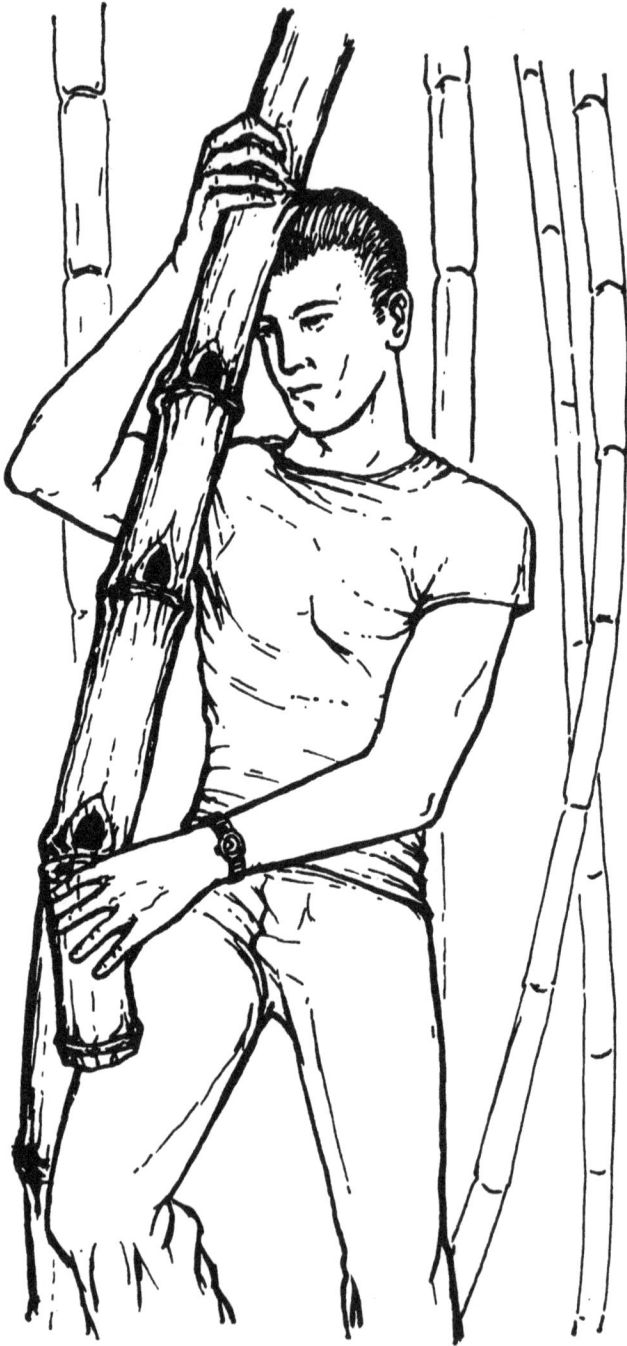

Figure 35. Bamboo joints contain water.

Figure 36.   Bromeliad catches water.

Figure 37. Traveler's tree.

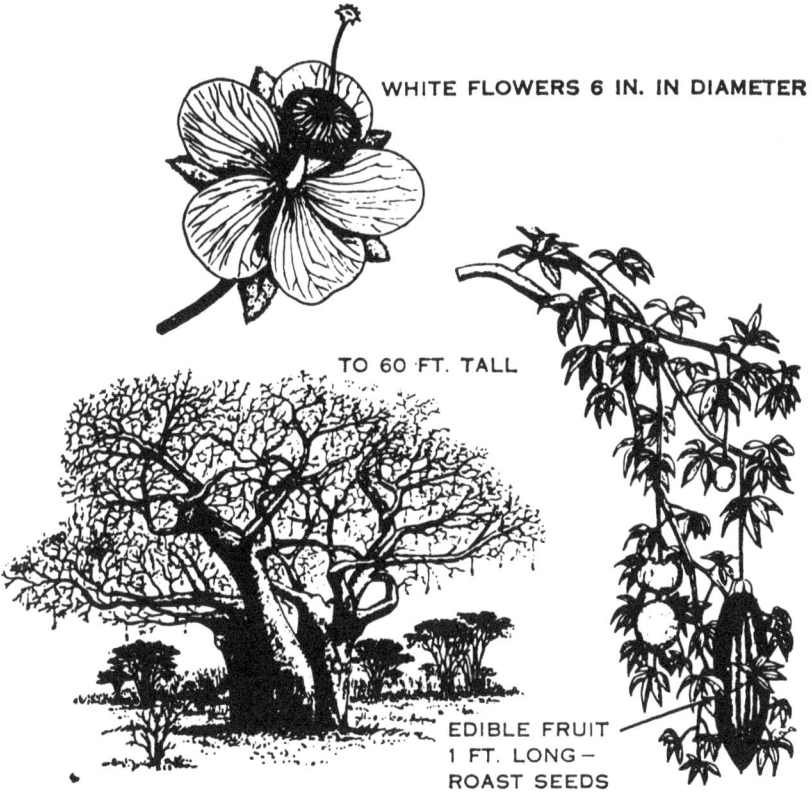

WHITE FLOWERS 6 IN. IN DIAMETER

TO 60 FT. TALL

EDIBLE FRUIT
1 FT. LONG –
ROAST SEEDS

Figure 38. Baobab tree.

CHAPTER 4

FOOD

Section I. GENERAL CONSIDERATIONS

32. You Needn't Starve

a. It takes little reasoning to recognize that "the second of things is food." This is especially true during a survival episode when you need every ounce of energy and endurance that you can muster.

b. Men have been known to live for more than a month without food, but unless you are in extreme circumstances, there is little need to be deprived of something to eat. Nature can be your provider if you know how to use her. Apply the following rules as soon as you realize that you are isolated:

(1) Inventory your rations and water. Estimate the length of time you will be on your own.

(2) Divide your food—two-thirds for the first half of your isolation and one-third for the second half.

(3) Avoid dry, starchy, and highly flavored foods and meats if you have less than one quart of water for each day. Remember—eating makes you thirsty. Eat food high in carbohydrates—hard candy, fruit bars.

(4) Keep strenuous work to a minimum. The less you work, the less food and water you require.

(5) Eat regularly if possible; don't nibble. Plan one good meal each day and cook it if you can. Cooking makes food safer, more digestible, and palatable. Also, the time you spend cooking will give you a rest period.

(6) Always be on the lookout for wild food. With few exceptions, everything you see that walks, crawls, swims, or grows from the soil is edible. Learn to live off the land.

Section II. VEGETABLE FOODS

33. General

a. Experts estimate that about 300,000 classified plants grow on the earth's surface, including many which grow on mountain tops and ocean floors. Of these, 120,000 varieties are edible. Obviously you won't be able to learn about all of these plants from reading this manual. If you know what to look for in the area in which you find yourself stranded, can identify it, and know how to prepare it properly, you should find enough to keep you alive. You may even surprise yourself with a delicious meal.

b. For the purposes of your study and later use, this manual gives descriptions and pictures of certain edible plants that you can eat. Become familiar with these "pilot plants"; they will enable you to evaluate the food possibilities of other plants of the same variety. To illustrate, the color of the juice of one plant might lead you to try another one in which the juice seems to be the same color and consistency.

c. Don't limit yourself to studying the illustrations and descriptions of plant food in this manual. Take every opportunity to see these plants in their natural habitat; then, if you are forced into a survival situation in any area of the world, you will know where the best plant foods of a region are.

d. Many of the edible plants you read about in this manual appear throughout the world. Blackberries and raspberries grow in the Philippines and Siberia as they do in America; cultivated potatoes, peas, and beans are found in Germany as well as in Canada; and persimmons thrive on Guam just as they do in Georgia.

e. Although plant food may not provide a balanced diet, especially in the Arctic where the heat producing qualities of meat are essential, it will sustain you. Many plant foods like nuts and seeds will give you enough protein for normal efficiency. In all cases, plants provide energy and calorie-giving carbohydrates.

f. Plants are available everywhere to provide the necessary energy while you forage for wild meat. You can depend on them to keep you alive if you are injured, unarmed in enemy territory, or in an area where wild life is not abundant.

34. Wild Plant Food

It is generally safe to try wild plant foods you see being eaten by birds and animals; however, you will find few plants of which every part is edible. Many have one or more identifiable parts that have considerable food—or thirst-quenching value. Here is a discussion of the plant parts that contain food value—

a. Roots and Other Underground Parts. These starch-storing foods include tubers, roots, root stalks, and bulbs.

(1) *Tubers.* All tubers are found below the ground and

must be dug. Cook them by boiling or roasting. See chapter 5.

(a) *Wild potato*. This is an example of an edible tuber.

EDIBLE TUBER

Figure 39. Wild potato.

The plant is small and found throughout the world, especially in the tropics (fig. 39).

(b) *Soloman's-seal.* Tubers of Soloman's-seal grow on small plants and are found in North America, Europe, Northern Asia, and Jamaica. Boiled or roasted, they taste much like parsnips (fig. 40).

(c) *Water chestnut.* The water chestnut is a native of Asia, but it has spread to both tropical and temperate areas of the world, including North America, Africa,

EDIBLE TUBER

Figure 40. Soloman's-seal.

and Australia. It is found as a free-floating plant on rivers, lakes, and ponds in quiet water. The plant covers large areas wherever it occurs and has two kinds of leaves—the submerged leaf, which is long, root-like, and feathery; and the floating leaves, which form a rosette on the surface of the water. The nuts borne beneath the water are an inch or two broad with strong spines that give them the appearance of a horned steer. The seed living within the horny structure may be roasted or boiled (fig. 41).

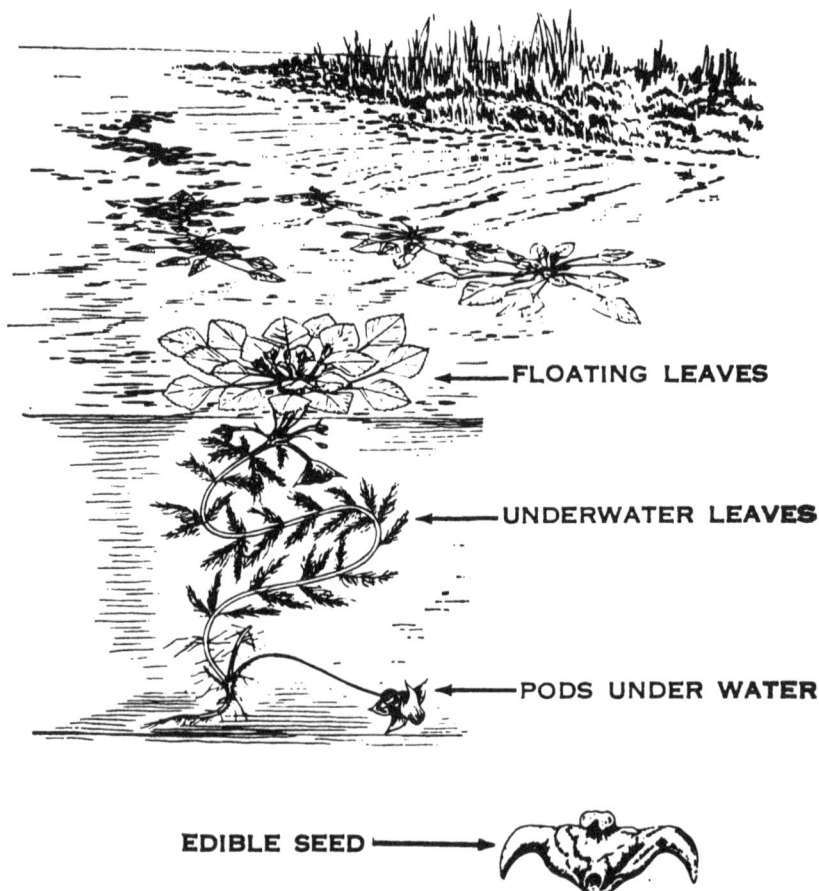

Figure 41. Water chestnut.

(d) *Nut grass.* Nut grass is widespread in many parts of the world. Look for it in moist sandy places along the margins of streams, ponds, and ditches. It occurs in both tropical and in temperate climates. The grass differs from true grass in that it has a three-angle stem and thick underground tubers that grow one-half to one inch in diameter. These tubers are sweet and nutty. Boil, peel, and grind them into flour. This flour can be used as a coffee substitute (fig. 42).

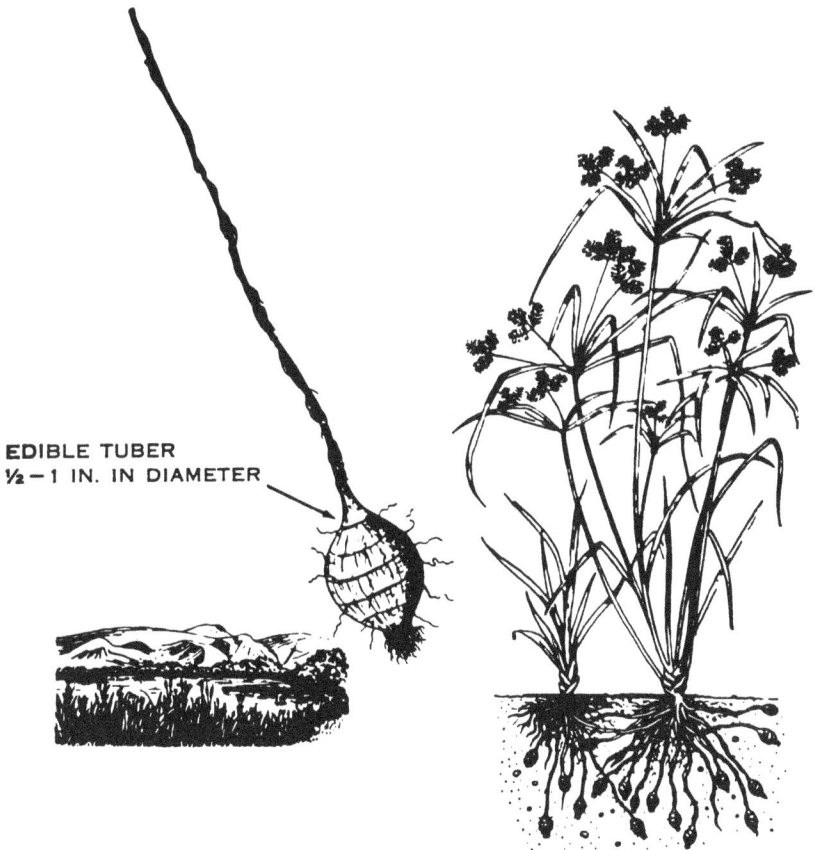

EDIBLE TUBER
½ – 1 IN. IN DIAMETER

Figure 42. Nut grass.

(e) *Taro.* The taro grows in moist, forested regions of nearly all tropical countries. It looks much like a calla lily and has leaves up to two feet long and stems five feet high. A pale yellow flower about 15 inches long blooms on this plant. It has an edible tuber growing slightly below ground level. This tuber must be boiled to destroy irritating crystals. After boiling, eat the tuber like a potato (fig. 43).

(2) *Roots and rootstalks.* These plant parts are storage devices rich in starch. Edible roots are often several

feet long and are not swollen like tubers. Rootstalks are underground stems, and some are several inches thick and relatively short and pointed. Following are illustrations showing both edible roots and rootstalks:

(a) *Bulrush.* This familiar tall plant is found in North America, Africa, Australia, East Indies, and Malaya. It is usually present in wet swampy areas. The roots and white stem base may be eaten cooked or raw (fig. 44).

(b) *Ti plant.* This plant is found in tropical climates, especially in the islands of the South Pacific. It is cultivated over wide areas of tropical Asia. In both the wild and cultivated state it ranges from 6 to 15 feet in height. It has large, coarse, shiny, leathery leaves arranged in crowded fashion at the tips of

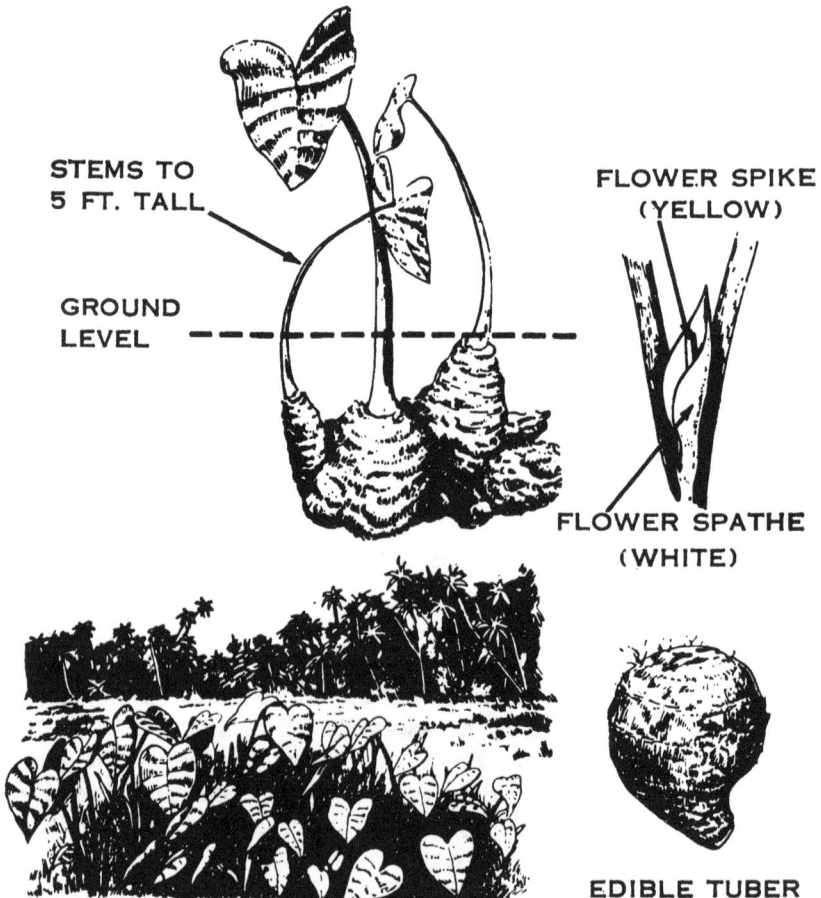

STEMS TO 5 FT. TALL

GROUND LEVEL

FLOWER SPIKE (YELLOW)

FLOWER SPATHE (WHITE)

EDIBLE TUBER

Figure 43. Taro.

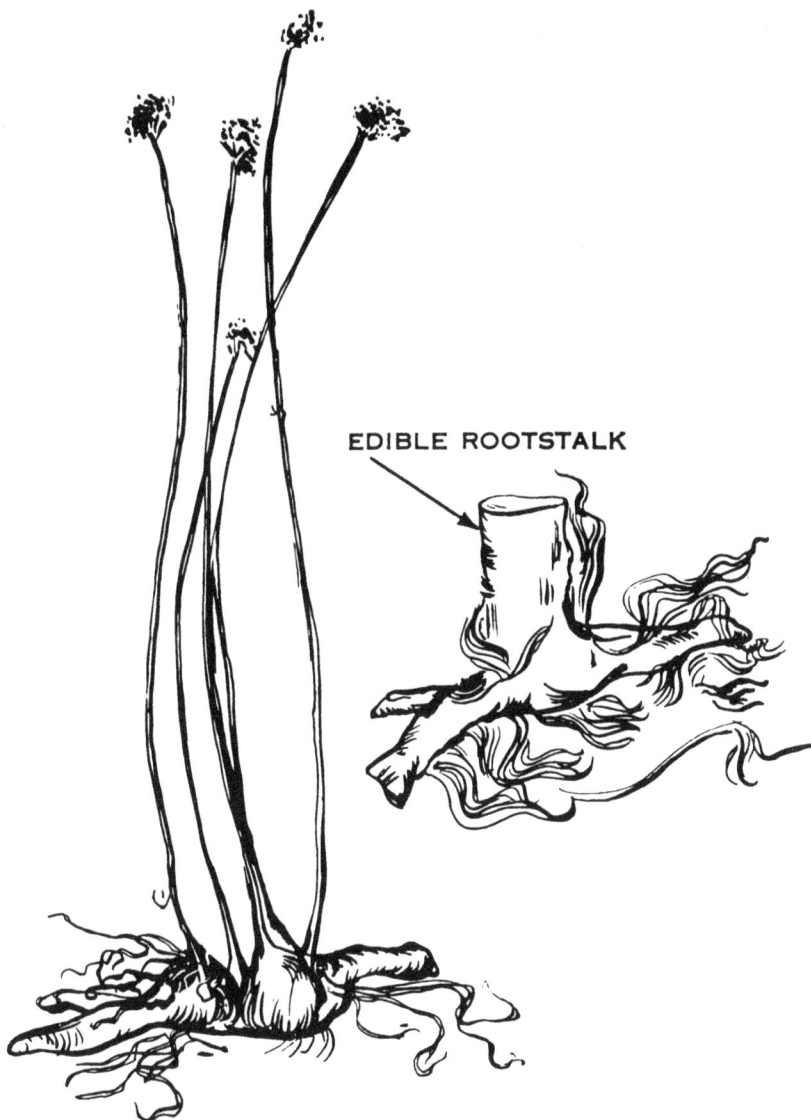

EDIBLE ROOTSTALK

Figure 44. Bulrush.

thick stems. The leaves are green and sometimes reddish. This plant grows a large plume-like cluster of
flowers that usually droop. It bears berries that are red
when ripe. The fleshy rootstalk is edible and full of
starch and should be baked for best results (fig. 45).

(c) *Water plantain.* This white flowered plant is found
most frequently around fresh water lakes, ponds, and
streams, where it is often partly submerged in a few

6-15 FT. TALL

EDIBLE
ROOTSTALK

Figure 45.   Ti plant.

inches of water. It is usually abundant in marshy
areas throughout the North Temperate Zone and has
long stalked, smooth, heart-shaped leaves with 3 to 9
parallel ribs. Thick, bulb-like rootstocks which grow
below the ground, lose their acrid taste after being
dried. Cook them like potatoes (fig. 46).

(d) *Flowering rush*. The flowering rush grows along
river banks, margins of lakes and ponds, and marshy
meadows over much of Europe and temperate Asia.
It grows in Russia and much of temperate Siberia.
The mature plant, usually found growing in a few
inches of water, reaches a height of three or more feet
and has loose clusters of rose-colored and green flowers.
The thick, fleshy underground rootstalk should be
peeled and boiled like potatoes (fig. 47).

(e) *Tapioca plant*. The tapioca or manioc plant is found
in all tropical climates, especially in wet areas. It
grows to a height of from 3 to 9 feet and has jointed

WHITE FLOWERS

EDIBLE STARCHY ROOTSTOCKS

Figure 46. Water plantain.

stems and finger-like leaves. There are two kinds of manioc that have edible rootstalks—bitter and sweet. The bitter manioc is the common variety in many areas and is poisonous unless cooked. If you find a rootstalk of bitter manioc, grind the root into a pulp and cook it for at least one hour. Flatten the wet pulp into cakes and bake. Another method of cooking this bitter variety is to cook the roots in large pieces for one hour then peel and grate them. Press the pulp and knead it with water to remove the milky juice.

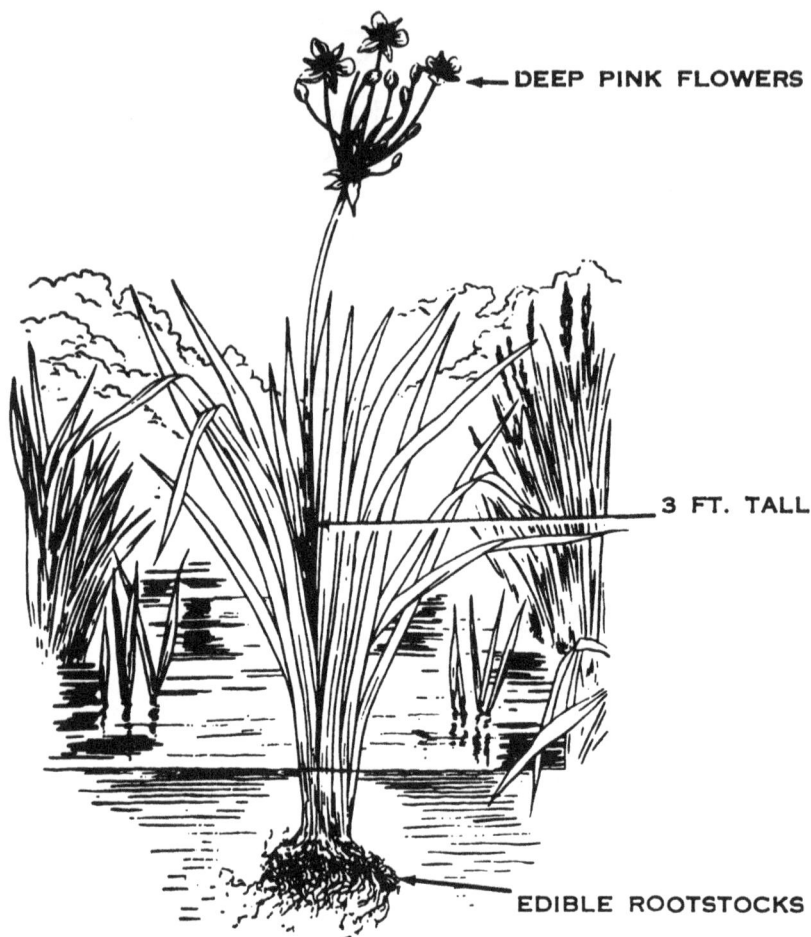

DEEP PINK FLOWERS

3 FT. TALL

EDIBLE ROOTSTOCKS

Figure 47. Flowering rush.

Steam it; then pour it into a plastic mass. Roll the paste into small balls and flatten them into thin cakes. Dry these cakes in the sun, and eat baked or roasted. Sweet manioc rootstalks are not bitter and can be eaten raw, roasted as a vegetable, or made into flour. You can use this flour to make dumplings or the cakes described above (fig. 48).

(f) *Cattail.* The cattail is found along lakes, ponds, and rivers throughout the world, except in the tundra and forested regions of the far north. It grows to a height of 6 to 15 feet with erect, still, tape-like, pale-green leaves one-quarter to one inch broad. Its edible rootstalk grows up to one inch thick and contains about 46 percent starch and 11 percent sugar. To prepare

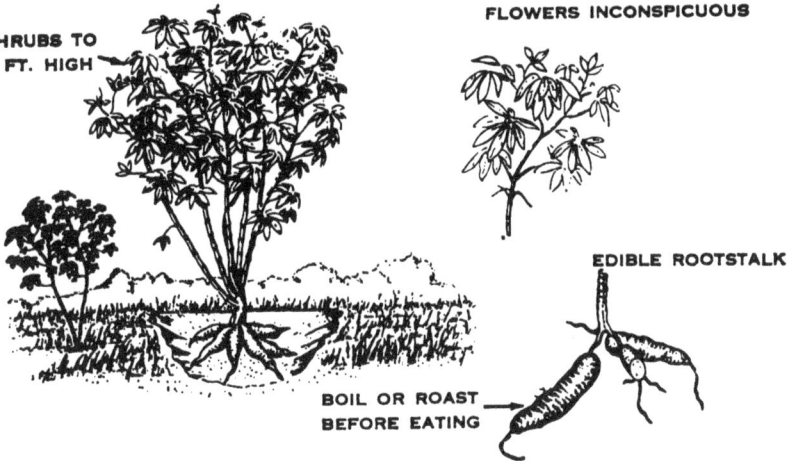

Figure 48. Tapioca.

these rootstalks, peel off the outer covering and grate the white inner portion. Eat them boiled or raw. The yellow pollen from the flowers can be mixed with water and steamed as bread. In addition, the young growing shoots are excellent when boiled like asparagus (fig. 49).

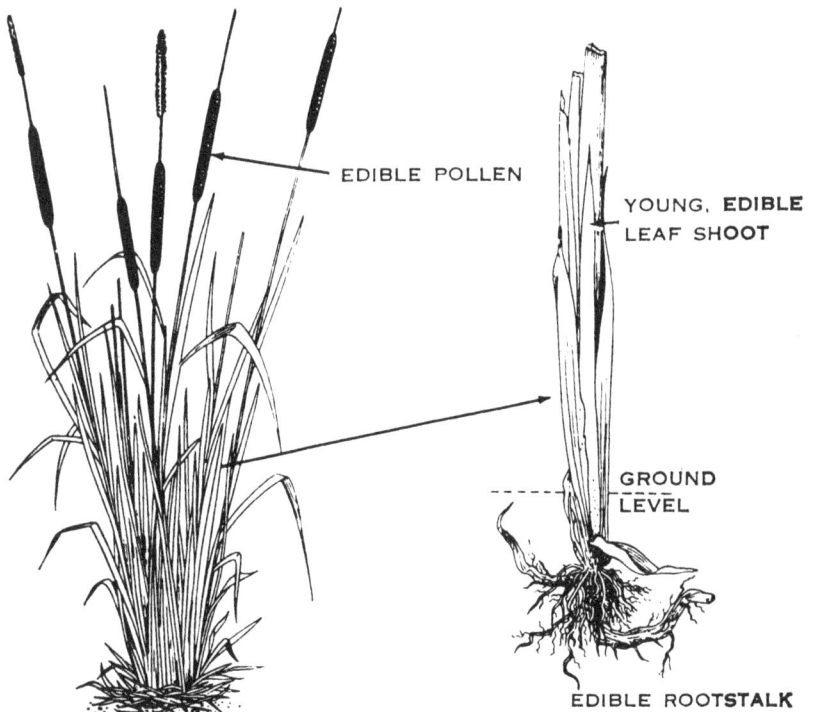

Figure 49. Cattail.

(3) *Bulbs.* All bulbs are high in starch content and, with the exception of the wild onion (*(a)* below), are more palatable if they are cooked.

  (a) *Wild onion.* This is the most common edible bulb and is a close relative to the cultivated onion. It is found throughout the North Temperate Zones of North America, Europe, and Asia. The plant grows from a bulb buried 3 to 10 inches below the ground. The leaves vary from grass-like to several inches wide. The plant grows a flower that may be white, blue, or a shade of red. No matter what variety of onion you find, you can detect it by its characteristic "oniony" odor. Eat the bulb. None is poisonous (fig. 50).

  (b) *Wild tulip.* The wild tulip is found in Asia Minor and Central Asia. The bulb of the plant can be cooked and eaten as a substitute for potatoes. The plant bears flowers for a short time in the spring and these

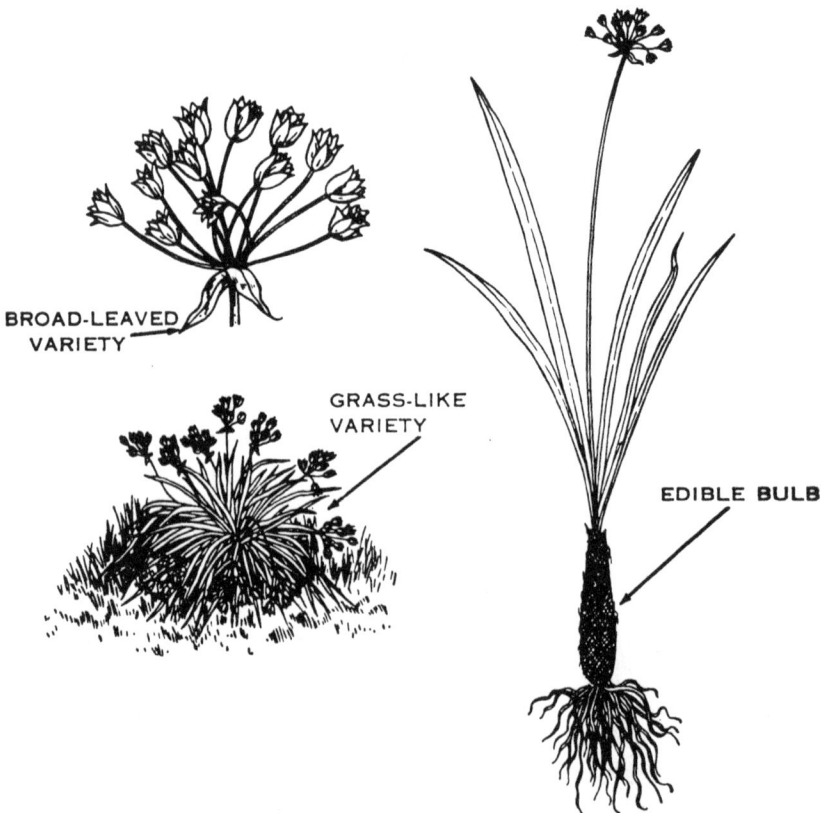

BROAD-LEAVED VARIETY

GRASS-LIKE VARIETY

EDIBLE BULB

Figure 50. Wild onion.

resemble the common garden tulip except that they are smaller. When red, yellow, or orange flowers are absent, a seed pod can be found as an identifying characteristic (fig. 51).

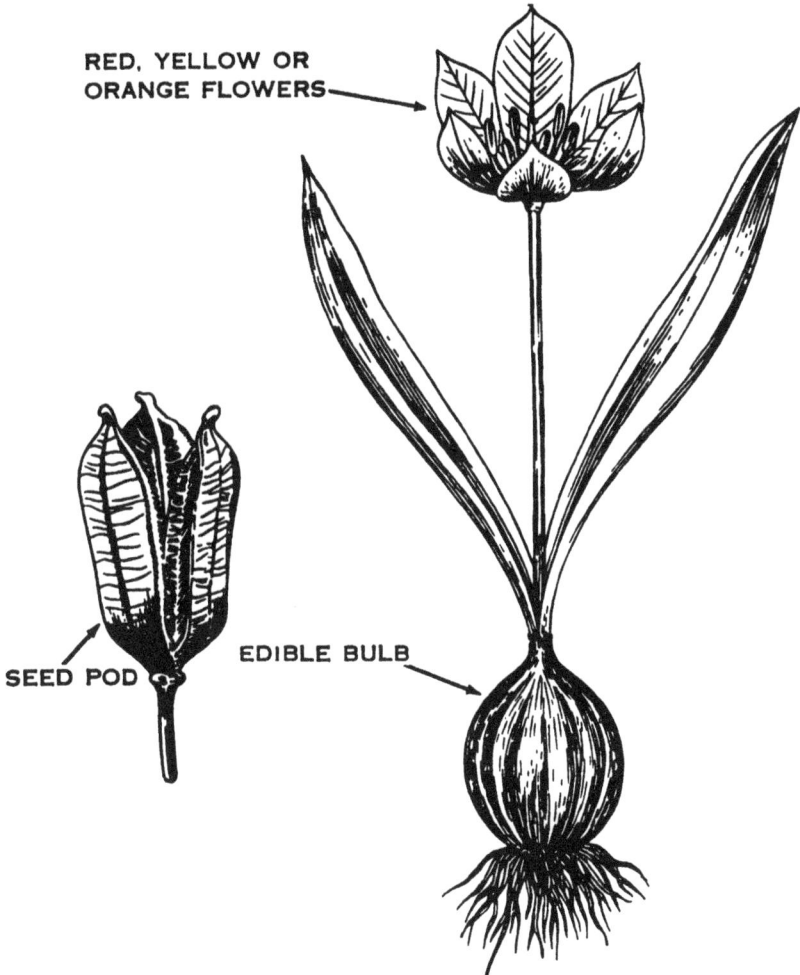

RED, YELLOW OR ORANGE FLOWERS

SEED POD

EDIBLE BULB

Figure 51. Wild tulip.

b. Shoots and Stems. Edible shoots grow very similarly to asparagus. The young shoots of ferns and bamboo, for example, make excellent food. Although some can be eaten uncooked, most shoots are better if they are parboiled for 10 minutes, the water drained off, and reboiled until they are sufficiently tender for eating. Here are a few of the plants you may find with edible shoots and stems—

(1) *Mescal.* This plant exists in Europe, Africa, Asia, Mexico, and the West Indies. It is a typical desert plant but also grows in moist tropical areas. The mescal, when fully grown, has thick, tough leaves with stout, sharp tips borne in a rosette. In the center is a stalk that rises like a candle to produce a flowering head. This stalk or shoot is the part to eat. Select plants having flowers not fully developed; roast the shoot. It contains fibrous, molasses-colored layers that taste sweet (fig. 52).

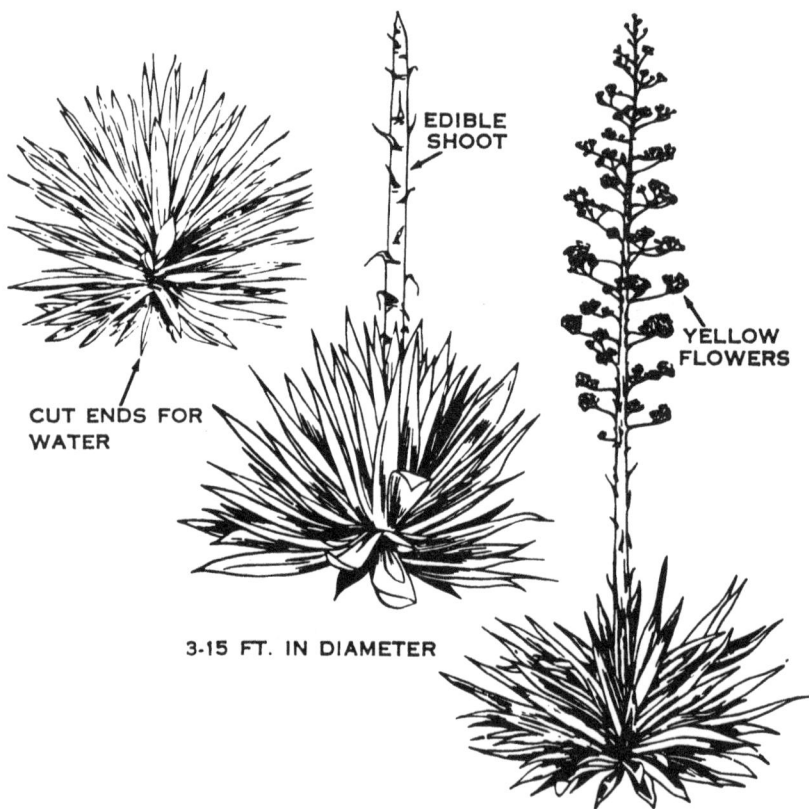

Figure 52. Mescal.

(2) *Wild gourd or luffa sponge.* This plant is a member of the squash family and grows similarly to watermelon, cantaloupe, and cucumber. It is widely cultivated in tropical areas, and it might be found in a wild state in old gardens or clearings. The vine has leaves 3 to 8 inches across and the fruit is cylindrical, smooth, and

seedy. Boil and eat the fruit when it is half ripe; make a meal on the tender shoots, flowers, and young leaves after cooking them. You can also roast the seeds and eat them like peanuts (fig. 53).

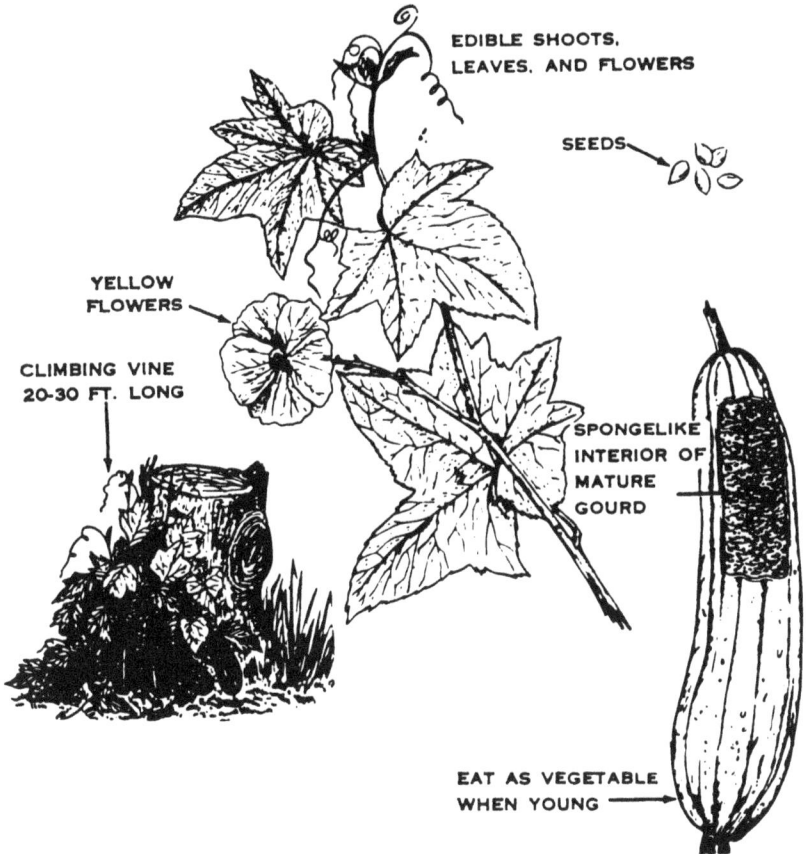

EDIBLE SHOOTS, LEAVES, AND FLOWERS

SEEDS

YELLOW FLOWERS

CLIMBING VINE 20-30 FT. LONG

SPONGELIKE INTERIOR OF MATURE GOURD

EAT AS VEGETABLE WHEN YOUNG

Figure 53.   Wild gourd.

(3) *Wild desert gourd.* Also a member of the squash family, this creeping plant grows abundantly in the Sahara desert, Arabia, and on the southeastern coast of India. It produces a vine 8 to 10 feet long that runs over the ground and a gourd that grows to about the size of an orange. The seeds are edible roasted or boiled. The flowers also can be eaten, and the water-filled stem shoots may be chewed (fig. 54).

(4) *Bamboo.* This plant grows in the moist areas of warm temperate and tropical zones. It is found in clearings, around abandoned gardens, in the forest, and along

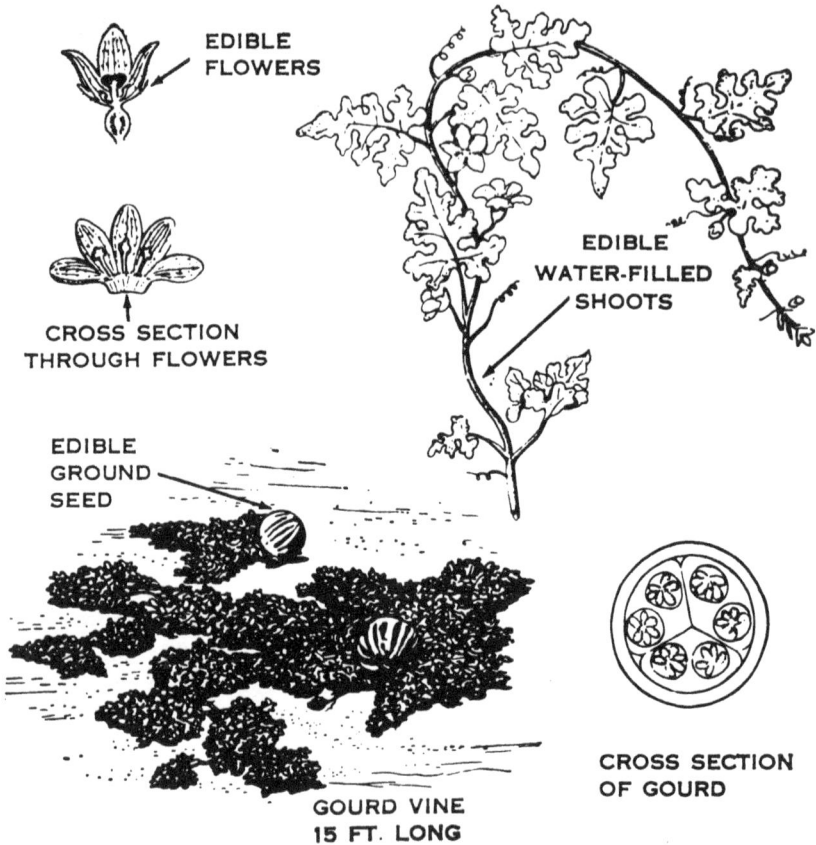

EDIBLE FLOWERS

CROSS SECTION THROUGH FLOWERS

EDIBLE WATER-FILLED SHOOTS

EDIBLE GROUND SEED

GOURD VINE 15 FT. LONG

CROSS SECTION OF GOURD

Figure 54. Desert gourd.

rivers and streams. Bamboo resembles corn plants and sugar cane and can be easily remembered for its popularity as fishing poles. The mature stems are very hard and woody, while the young shoots are tender and succulent. Cut these young shoots the same as you would cut asparagus, and eat the soft tip ends after boiling. Freshly cut shoots are bitter, but a second change of water eliminates the bitterness. Remove the tough protective sheath around the shoot before you eat it. Also edible is the seed grain of the flowering bamboo. Pulverize this, add water, and press it into cakes or boil it as you would rice (fig. 55).

(5) *Edible ferns.* Ferns are abundant in moist areas of all climates, especially in forested areas, gullies, along streams, and on the edge of woods. You might mistake them for flowering plants, but by using a little care you will be able to distinguish them from all other green

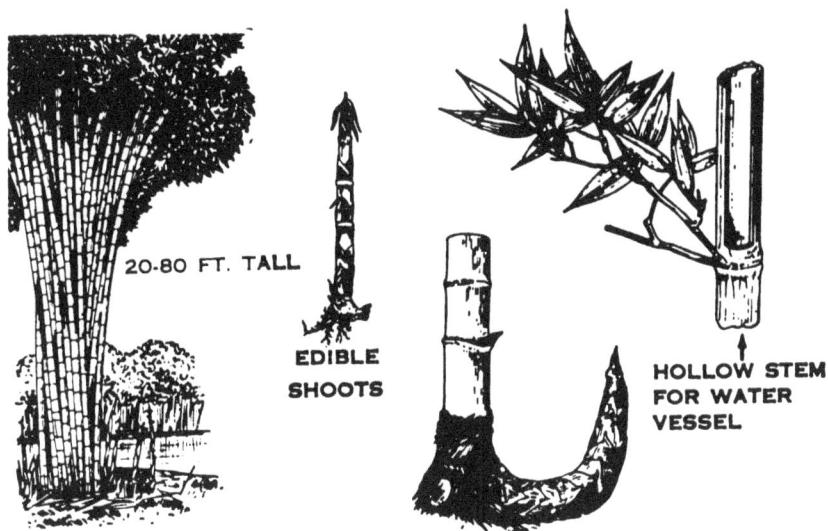

20-80 FT. TALL

EDIBLE
SHOOTS

HOLLOW STEM
FOR WATER
VESSEL

Figure 55. Bamboo.

plants. The under surface of the leaves is usually covered with masses of brown dots, which themselves are covered with yellow, brown, or black dust. These dots are filled with spores and their presence makes them easily distinguishable from plants with flowers.

(a) Bracken is one of the most widely distributed ferns. It occurs throughout the world, except the Arctic, in open, dry woods, recently burned clearings, and pastures. It is a course fern with solitary or scattered young stalks, often one-half inch thick at the base, nearly cylindrical, and covered with rusty felt; the uncoiling frond is distinctly three-forked with a purplish spot at each angle. This spot secretes a sweetish juice. Old fronds are conspicuously three-forked, and the rootstock is about one-quarter inch thick, creeping, branching, and woody (fig. 56).

(b) On all ferns, select young stalks (fiddleheads) not more than 6 to 8 inches high. Break them off as low as they remain tender; then close your hand over the stalk and draw it through to remove the wool. Wash and boil in salted water or steam until tender (fig. 57).

c. Leaves. Plants which produce edible leaves are probably the most numerous of all plant foods. You can eat many of them raw or cooked; however, if you cook them don't cook them too long as many of the valuable vitamins will be destroyed. Following are some plants with edible leaves:

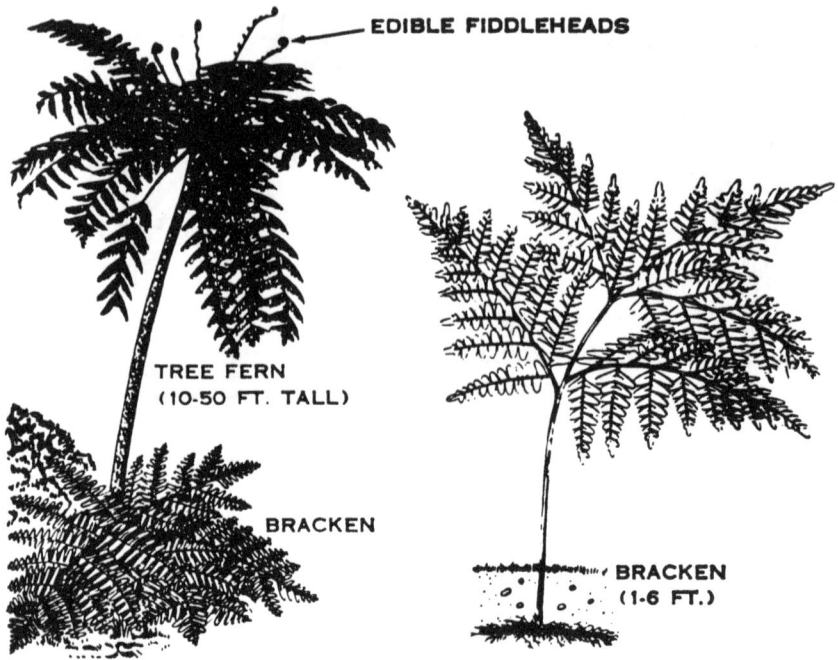

EDIBLE FIDDLEHEADS

TREE FERN
(10-50 FT. TALL)

BRACKEN

BRACKEN
(1-6 FT.)

Figure 56.   Bracken.

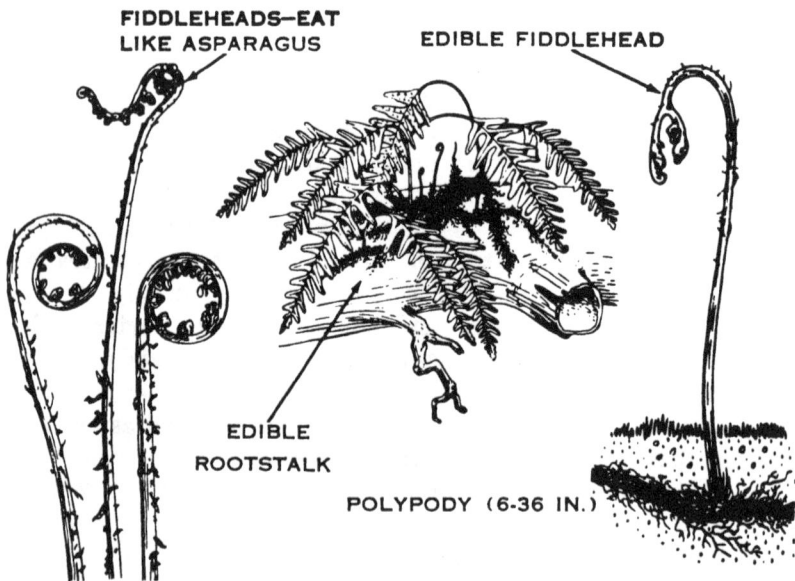

FIDDLEHEADS—EAT
LIKE ASPARAGUS

EDIBLE FIDDLEHEAD

EDIBLE
ROOTSTALK

POLYPODY (6-36 IN.)

EDIBLE FERNS

Figure 57.   Edible fern parts.

(1) *Baobab*. This tree is found in open bush country throughout tropical Africa. You can spot it by its enormous girth and swollen trunk and the relatively low stature of the tree. A mature tree 60 feet high may have a trunk 30 feet in diameter. It grows large white flowers about three inches across that hang loosely from the tree. The tree also bears a mealy pulpy fruit with numerous seeds. These are edible and you can use the leaves as a soup vegetable (fig. 38).

(2) *Ti plant*. See *a* (2) *(b)* above.

(3) *Water lettuce*. This plant grows throughout the Old World tropics in both Africa and Asia and in the New World tropics from Florida to South America. It is found only in very wet places, usually as a floating water plant. Look for it in still lakes, ponds, and backwaters. Look for little plantlets growing from the margins of the leaves. These grow in the shape of a rosette, and they often cover large areas in the regions in which they are found. The plant's leaves look much like lettuce and are very tender. Eat the fresh leaves like lettuce but eat only those that are well out of the water (fig. 58).

(4) *Spreading wood fern*. This plant is especially abundant in Alaska and Siberia and is found in mountains and woodlands. It sprouts from stout underground stems which are covered with old leafstalks that resemble a bunch of small bananas. Roast these leafstalks and remove the shiny brown covering. Eat the inner portion. In the early spring, collect the young fronds or fiddleheads, boil or steam them, and eat them like asparagus (fig. 59).

(5) *Horse-radish tree*. This tropical plant is native to India but is widespread in other tropical countries throughout southern Asia, Africa, and America. Look in abandoned fields and gardens and on the edge of forest areas for a rather low tree from 15 to 45 feet high. The leaves have a fern-like appearance and can be eaten old or young, fresh or cooked, depending on their state of hardness. At the ends of the branches are flowers and long pendulous fruit that resemble a giant bean. Cut the long, young seed pod into short lengths and cook it like string beans. You can chew young seed pods when they are fresh. The roots of this plant are pungent and you can grind them for seasoning much as you do true horse-radish (fig. 60).

MATURE PLANT
(2-3 FT. TALL)

EDIBLE YOUNG
FLOATING LEAVES

Figure 58. Water lettuce.

(6) *Wild dock and wild sorrel.* Although these plants are
native to the Middle East they are often abundant in
both temperate and tropical countries and in areas of
high as well as low rainfall. Look for them in fields,
along roadsides, and in waste places. Wild dock is a
stout plant with most of its leaves at the base of its
6- to 12-inch stem. It produces a very small, green to
purplish plume-like cluster of flowers. Wild sorrel is
smaller than dock but similar in appearance. Many of
its basal leaves are arrow-shaped and contain a sour
juice. The leaves of both plants are tender and you can
eat them fresh or slightly cooked. To take away the
strong taste, change the water once or twice during the
cooking (fig. 61).

(7) *Wild chicory.* Originally a native of Europe and Asia,
chicory is now generally distributed throughout the

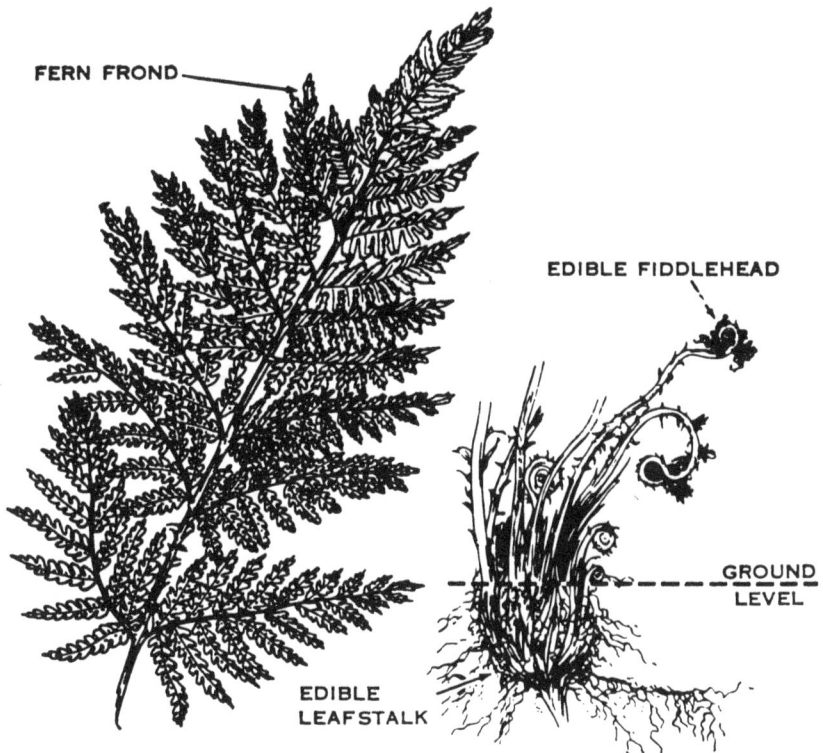

Figure 59. Spreading wood fern.

United States as a weed along roadsides and in fields. Its leaves are clustered at ground level at the top of a strong underground carrot-like root. The leaves look much like dandelion but are thicker and rougher. The stems rise 2 to 4 feet and are covered in summer with numerous bright blue heads of flowers, also resembling a dandelion except for color. You can eat the tender young leaves as a salad without cooking and grind the roots as a coffee substitute (fig. 62).

(8) *Arctic willow.* This shrub never exceeds 1 or 2 feet in height and is common on all tundra areas in North America, Europe, and Asia. It grows in clumps which form dense mats on the tundra. You can collect young shoots in the early spring and eat the inner portion raw after stripping off the outer bark. The young leaves are a rich source of vitamin C containing 7 to 10 times more than oranges (fig. 63).

(9) *Lotus lily.* This plant grows in fresh water lakes, ponds, and slow streams from the Nile basin through Asia to

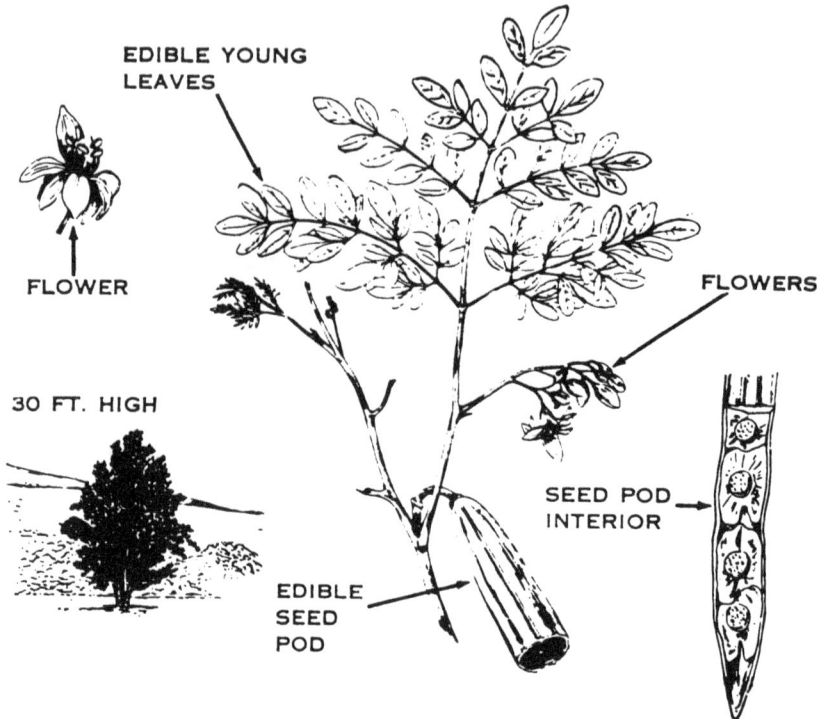

Figure 60. Horse-radish tree.

China and Japan and southward to India. It also grows in the Philippines, Indonesia, northern Australia, and eastern United States. The leaves of the lotus lily are shield-shaped, 1 to 3 feet across. They stand 5 to 6 feet above the surface of the water and grow either pink, white, or yellow flowers 4 to 10 inches in diameter. Eat the young stems and leaves after you cook them but remove the rough, outer layer of the young stems before cooking or eating. The seeds are also edible when ripe. Remove the bitter embryo, then boil or roast them. Also edible are the rootstalks, which become 50 feet long with tuberous enlargements. Boil these and eat them like potatoes (fig. 64).

(10) *Papaya.* This tree grows in all tropical countries, especially in moist areas. You will find it around clearings and former habitations, and also in open sunny places in uninhabited jungle areas. The papaya tree is small (6 to 20 feet tall) with a soft hollow trunk that will break under your weight if you try to climb it. This trunk is rough and the leaves are crowded at the top.

Figure 61. Wild dock and wild sorrel.

The yellow or greenish fruit grows among and below the leaves directly from the trunk and is squash-shaped. It is high in vitamin C and you can eat it raw or cooked. The milky sap of unripe fruit is a good meat tenderizer if you rub it into the meat. Be careful not to get this juice into your eyes—it will cause intense pain and temporary or even permanent blindness. Also edible are the young papaya leaves, flowers, and stems. Cook them carefully and change the water at least twice (fig. 65).

(11) *Wild rhubarb.* This plant grows from southeastern Europe to Asia Minor through the mountainous regions of central Asia to China and can be found in open places, along the borders of woods and streams, and on mountain slopes. The large leaves grow from the base of long stout stalks. These stalks flower and rise above the large leaves and may be boiled and eaten as vegetable (fig. 66).

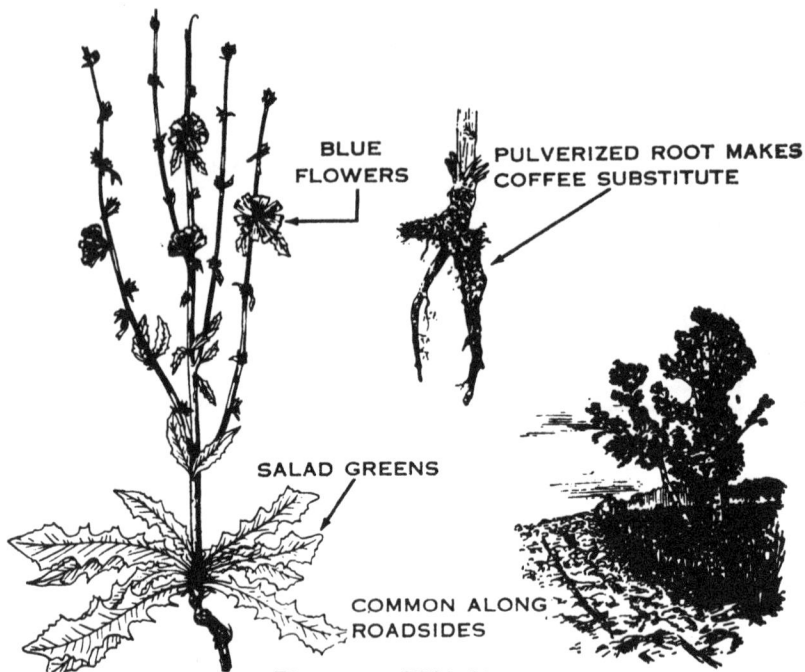

BLUE FLOWERS

PULVERIZED ROOT MAKES COFFEE SUBSTITUTE

SALAD GREENS

COMMON ALONG ROADSIDES

Figure 62. Wild chicory.

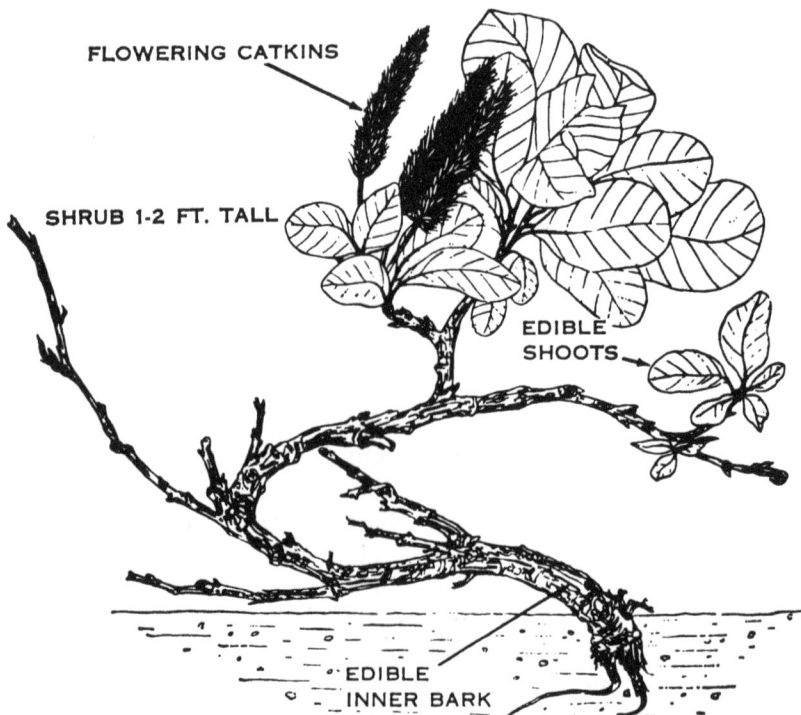

FLOWERING CATKINS

SHRUB 1-2 FT. TALL

EDIBLE SHOOTS

EDIBLE INNER BARK

Figure 63. Arctic willow.

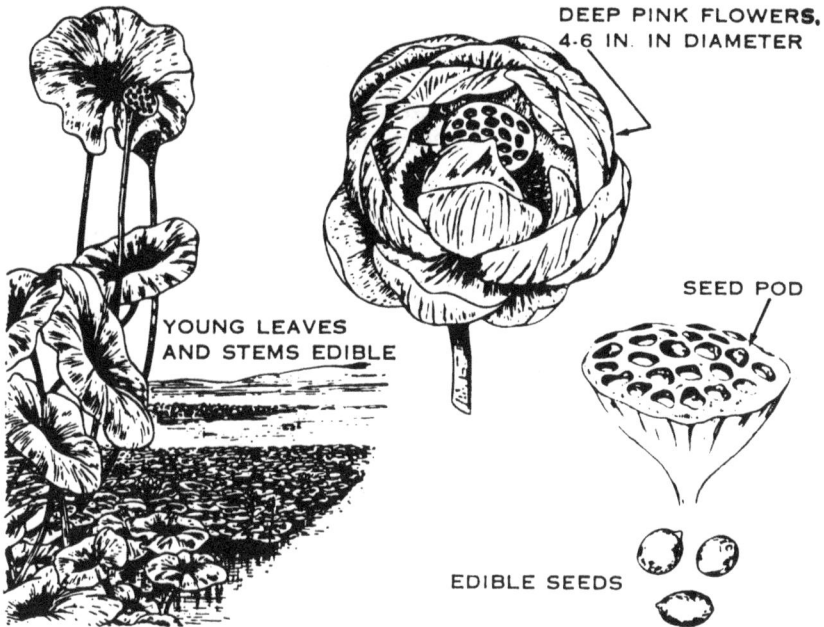

DEEP PINK FLOWERS,
4-6 IN. IN DIAMETER

SEED POD

YOUNG LEAVES
AND STEMS EDIBLE

EDIBLE SEEDS

Figure 64. Lotus lily.

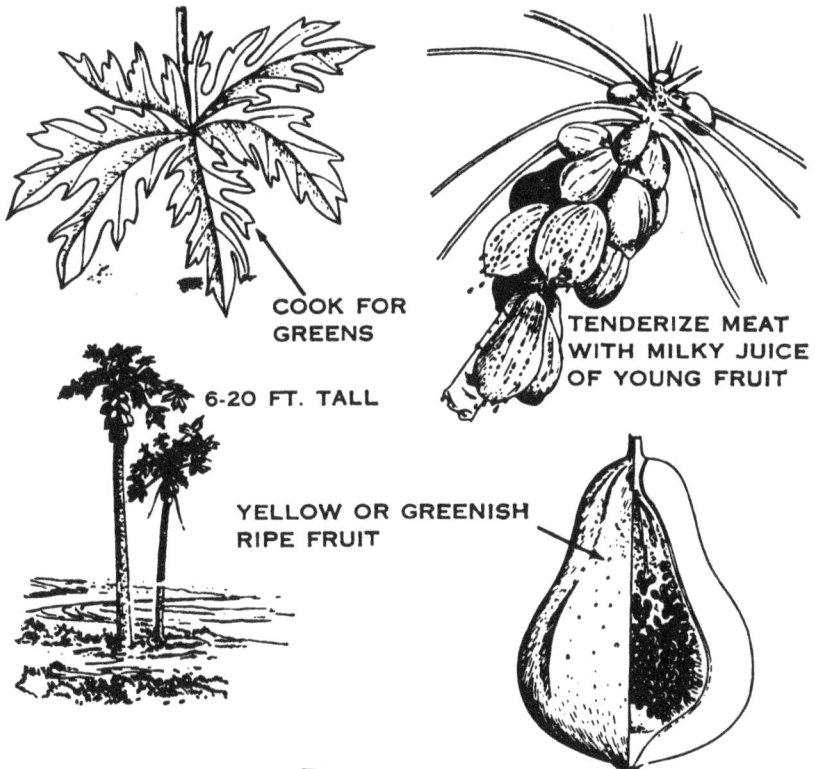

COOK FOR
GREENS

6-20 FT. TALL

TENDERIZE MEAT
WITH MILKY JUICE
OF YOUNG FRUIT

YELLOW OR GREENISH
RIPE FRUIT

Figure 65. Papaya.

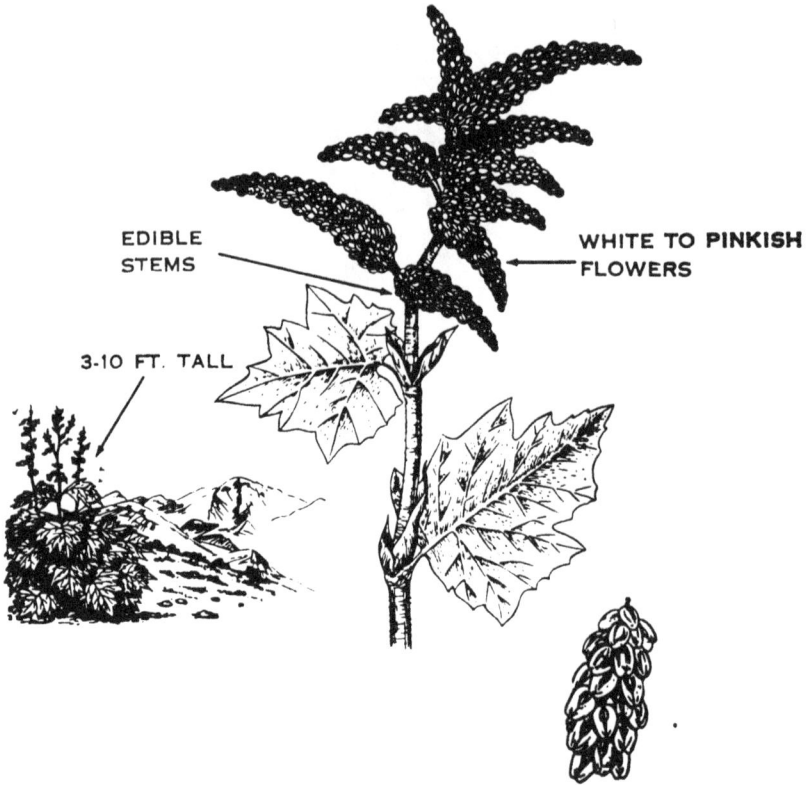

EDIBLE STEMS

WHITE TO **PINKISH** FLOWERS

3-10 FT. TALL

Figure 66. Wild rhubarb.

(12) *Prickly pear.* This plant is native to America, but grows in many desert areas of the Old World and Australia. It is found in the southwestern United States, Mexico, South America, and along the shores of the Mediterranean. It has a thickened stem about an inch in diameter which is full of water. The outside is covered with clusters of very sharp spines spaced at intervals, and the plant grows yellow or red flowers. Do not mistake this plant for other kinds of thick, fleshy cactus-like plants, especially those in Africa. The spurges of Africa look like cactuses but contain a milky, poisonous juice. The prickly pear never gives a milky juice. You can eat the egg-shaped fruit growing at the top of the cactus. Slice off the top of the fruit, peel back the outer layer, and eat the contents, seeds and all. Also edible are the prickly pear pads. Cut away the spines and slice the pad lengthwise into strips like string beans. Eat them raw or boiled (fig. 67).

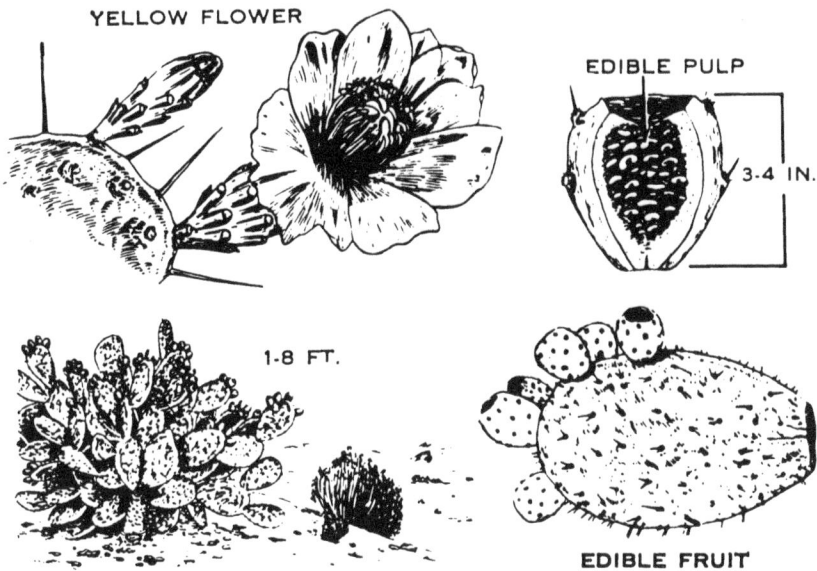

YELLOW FLOWER

EDIBLE PULP

3-4 IN.

1-8 FT.

EDIBLE FRUIT

Figure 67. Prickly pear.

d. Nuts. Nuts are among the most nutritious of all plant foods and contain valuable protein. Plants bearing edible nuts grow in all climatic zones and continents of the world except the Arctic. You are probably familiar with nuts of the temperate zones such as walnuts, filberts, almonds, hickory nuts, acorns, hazelnuts, beechnuts, and pine nuts. Tropical zone nuts include coconuts, Brazil nuts, cashew nuts, and macademia nuts. Following are some examples of nuts that you can eat:

(1) *English walnut.* In the wild state this nut is found from Southeastern Europe across Asia to China. It is abundant in the Himalayas and grows on a tree that sometimes reaches 60 feet tall. The leaves of the tree are divided and are characteristic of all walnut species. The walnut itself is enclosed by a thick outer husk which must be removed to reach the hard inner shell of the nut. The nut kernel ripens in autumn (fig. 68).

(2) *Hazelnut* (Filbert). Hazelnuts are found over wide areas of the United States, especially the eastern half of the country. They grow also in Europe and eastern Asia from the Himalayas to China and Japan. Growing mostly on bushes 6 to 12 feet tall, hazelnuts exist in dense thickets along stream banks and open places. The nut is enveloped by a bristly long-necked husk; it ripens in the fall. You can eat it either in the dried or fresh

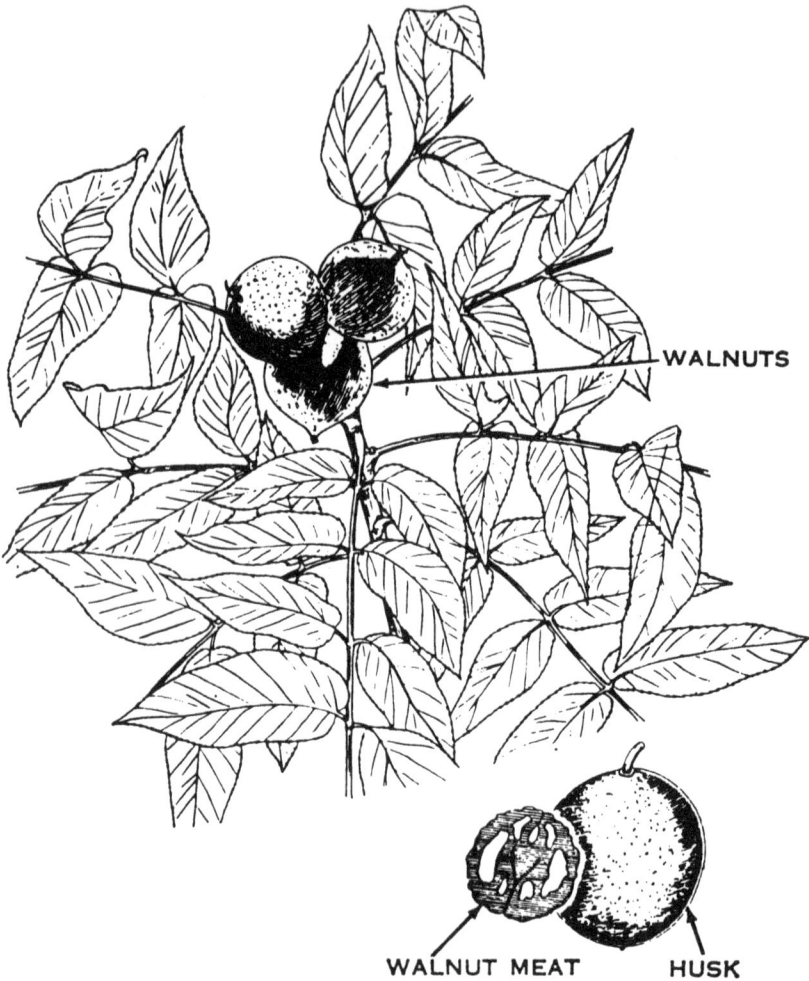

Figure 68. Walnut.

WALNUTS

WALNUT MEAT   HUSK

unripe stage, and derive great food value from its oil content (fig. 69).

(3) *Chestnut*. Wild chestnuts are highly useful as a survival food. They grow in central and southern Europe, from central Asia to China and Japan. The European chestnut is the most common variety; it grows along the edge of meadows and is a forest tree some 60 feet in height. You can prepare either the ripe or unripe nut by roasting it in embers or by boiling the kernel that lies within the shell. If you boil the nut, mash it like potatoes before eating it (fig. 70).

(4) *Almond*. Wild almonds grow in the semidesert areas of

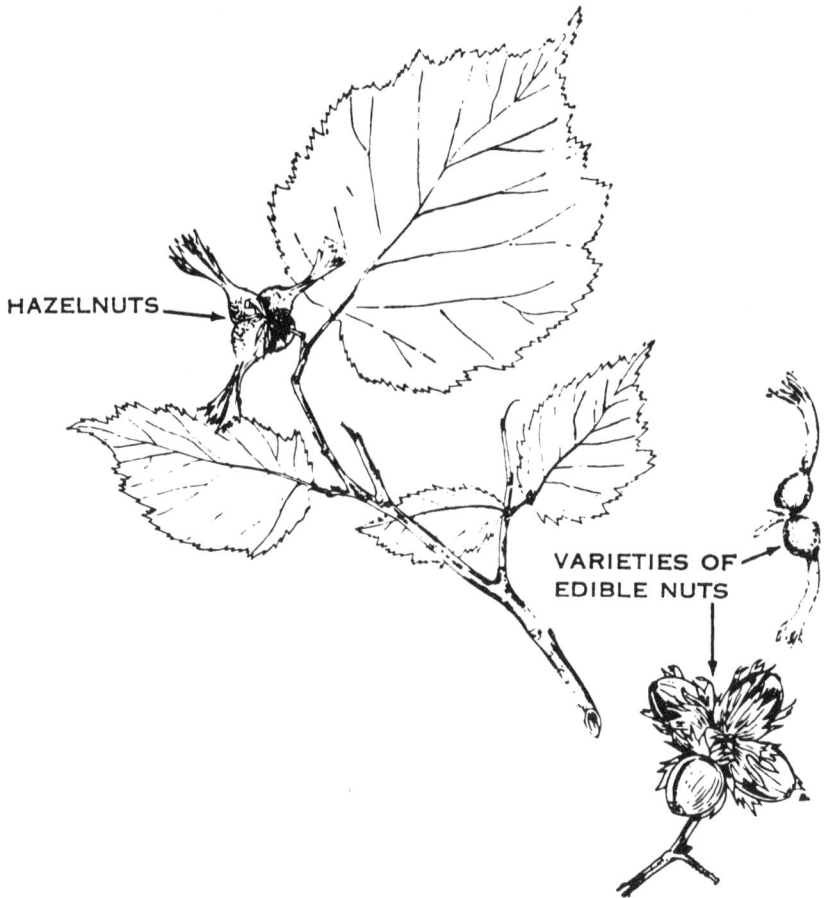

HAZELNUTS

VARIETIES OF EDIBLE NUTS

Figure 69. Hazelnut.

southern Europe, the eastern Mediterranean area, Iran, Arabia, China, Madeira, the Azores and the Canary Islands. The almond tree resembles a peach tree and sometimes grows 40 feet tall. The fruit, found in clusters all over the tree, looks somewhat like a gnarled, unripened peach with its stone (the almond itself) covered with a thick, dry wooly skin. To get at the almond nut, split the fruit down the side, then crush open the hard stone. It is a good idea to gather and shell them in large quantities for further use as a survival food because you can live for a long period on nothing but almonds (fig. 71).

(5) *Acorns* (English Oak). There are many varieties of oak, but the English oak is typical of those found in the North Temperate Zone. It often grows 60 feet tall and

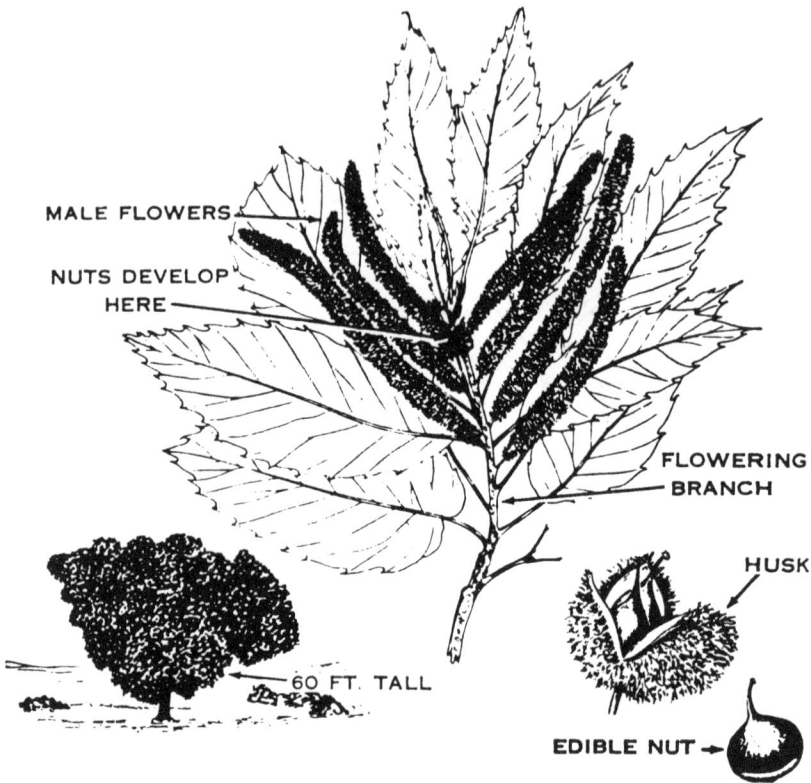

MALE FLOWERS

NUTS DEVELOP HERE

FLOWERING BRANCH

HUSK

60 FT. TALL

EDIBLE NUT

Figure 70. Chestnut.

the leaves are deeply lobed. The acorns grow out of a cup and are not edible raw because of the bitter tannin properties of the kernel. Boil the acorns for two hours, pour off the water, and soak the nut in cold water. Change the water occasionally and after 3 to 4 days, grind the acorns into paste. Make the paste into mush by mixing it with water and cooking it. You can make flour out of this paste by spreading and drying it (fig. 72).

(6) *Beechnut.* Beechnut trees grow wild in moist areas of the eastern United States, Europe, Asia, and North Africa. They are common throughout southeastern Europe and across temperate Asia but do not grow in tropical or subarctic areas. The beechnut is a large tree, sometimes reaching 80 feet in height, with smooth, light-gray bark, and dark green foliage. Mature beechnuts fall out of their husk-like seed pods, and you can get to the nut by breaking the thin shell with your finger nail. Roast and pulverize the kernel; then boil

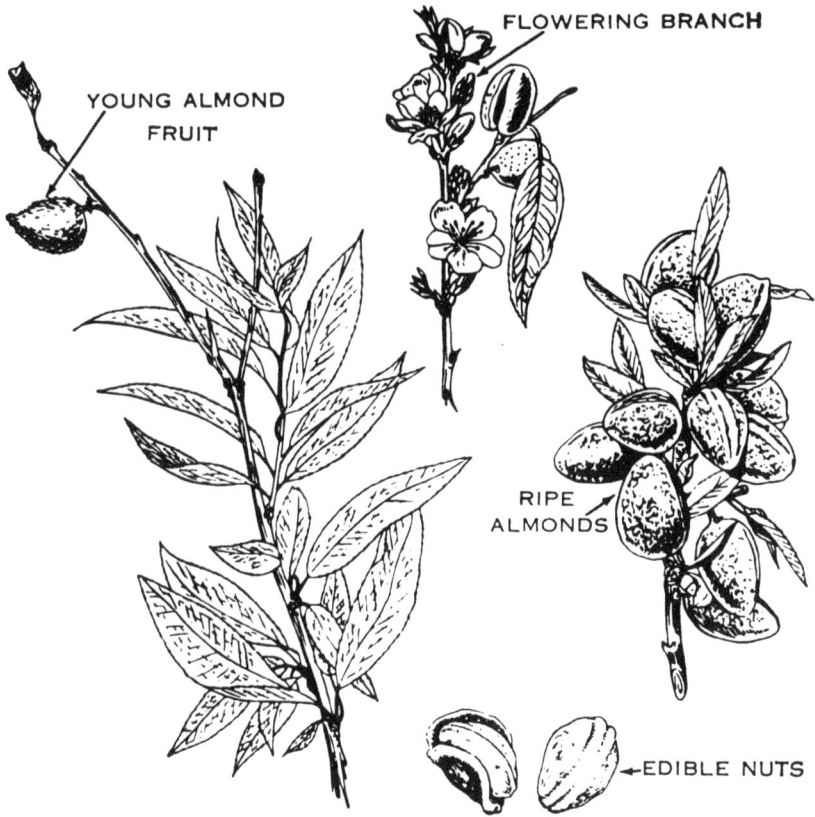

Figure 71. Almonds.

the powder for a passable coffee substitute (fig. 73).

(7) *Swiss stone pine.* Although it is unnecessary to distinguish between pines in order to use the pine nut, the Swiss pine shown in figure 74 will give you an idea of what this tree has to offer. Swiss stone pine is distributed widely in Europe and northern Siberia. Like all pines it is evergreen with needles growing in bunches. The edible seeds or nuts grow in woody cones which hang either separately or in clusters near the tips of the branches. The nuts grow at the base of the cone scales and when mature will fall out of the ripe cone. Eat these raw or roasted.

(8) *Water chestnut.* (a(1)(c) above.)

(9) *Tropical almond.* The Indian or tropical almond tree is widely dispersed in all tropical countries and is found in abandoned fields, gardens, along roadsides, and upon sandy seacoasts. It sometimes grows 100 feet tall. The

LARGE OAK

ACORNS

EDIBLE
ACORNS

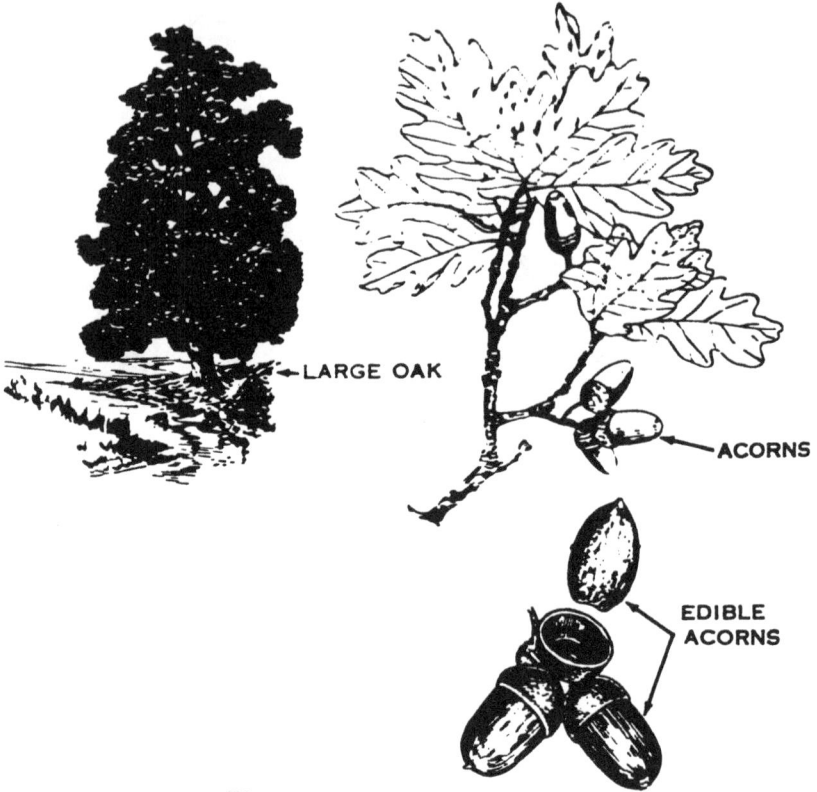

Figure 72.   Acorns (English Oak).

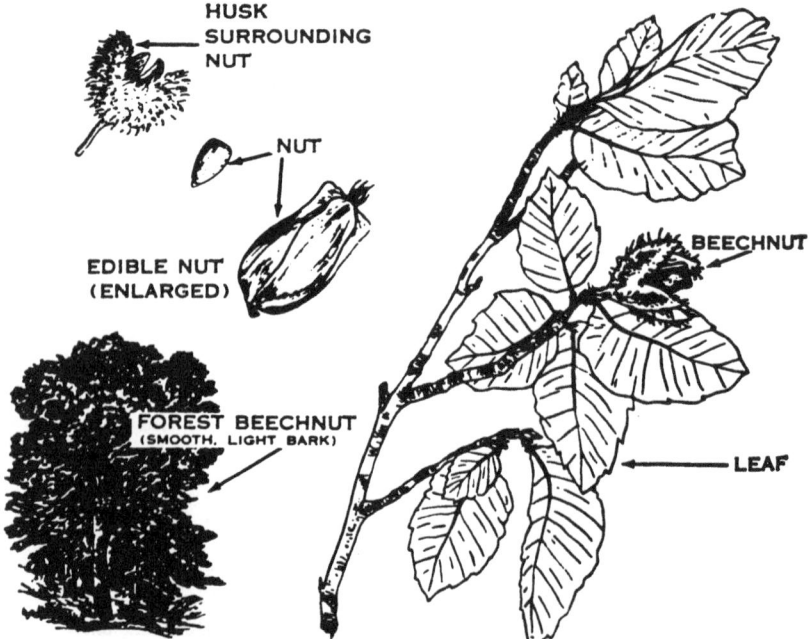

HUSK
SURROUNDING
NUT

NUT

EDIBLE NUT
(ENLARGED)

BEECHNUT

FOREST BEECHNUT
(SMOOTH, LIGHT BARK)

LEAF

Figure 73.   Beechnut tree.

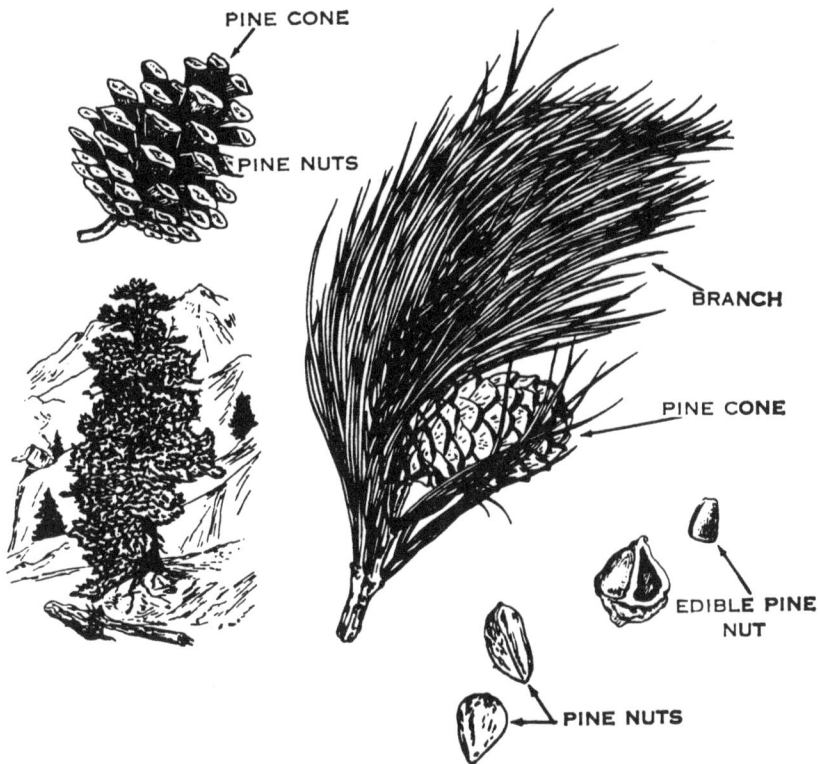

Figure 74. Pine nuts.

edible seeds or kernels growing at the tips of the branches are surrounded by a spongy, husklike covering from 1 to 3 inches long. These kernels have an almond-like consistency and flavor (fig. 75).

(10) *Coconut.*

    (a) The coconut palm is widely cultivated but grows wild throughout much of the moist tropics. It exists mainly near the seashore, but it sometimes grows some distance inland. This tall, unbranched tree sometimes reaches 90 feet. The nuts grow in large clusters and hang downward among the leaves.

    (b) The two most valuable parts of the coconut palm are the cabbage and the nut. The cabbage is the snow-white heart at the top of the tree. Eat it cooked, raw, or mixed with vegetables. The nut is most useful in the drinking and mature stage. In the drinking stage split the nut and scoop out the meat with a spoon fashioned from the outside husk. In the mature stage, crack the nut, loosen the meat, and eat it fresh,

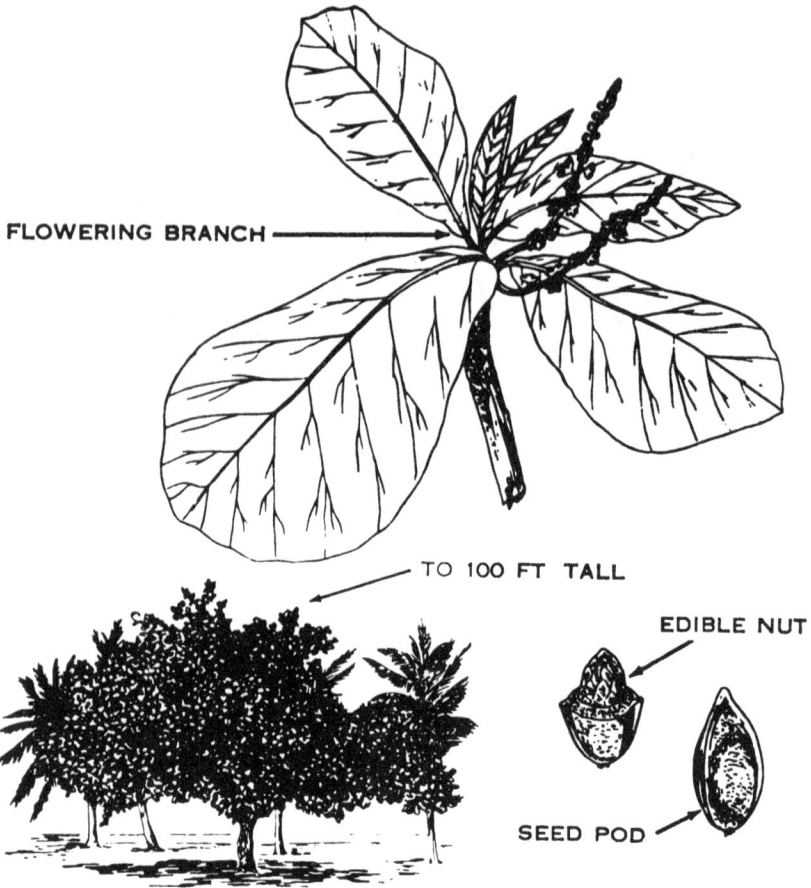

FLOWERING BRANCH

TO 100 FT TALL

EDIBLE NUT

SEED POD

Figure 75. Tropical almond.

grated, or dried to copra. Let the milk stand for a
short time so that the oil will separate from it, and
you will be able to use it for food or drink. See
figure 31.

(c) You can also eat sprouting coconuts. Husk and split
them open or simply crack them in half. Eat the
white spongy material inside. To remove the purgative or physic qualities of this meat cook it before
eating (fig. 76).

(11) *Wild pistachio nut.* About seven types of wild pistachio nuts occur in desert or semidesert areas surrounding
the Mediterranean, in Asia Minor, and Afghanistan.
Some are evergreen while others lose their leaves during
the dry season. The leaves alternate on the stem and
have either three large leaves or a number of leaflets.

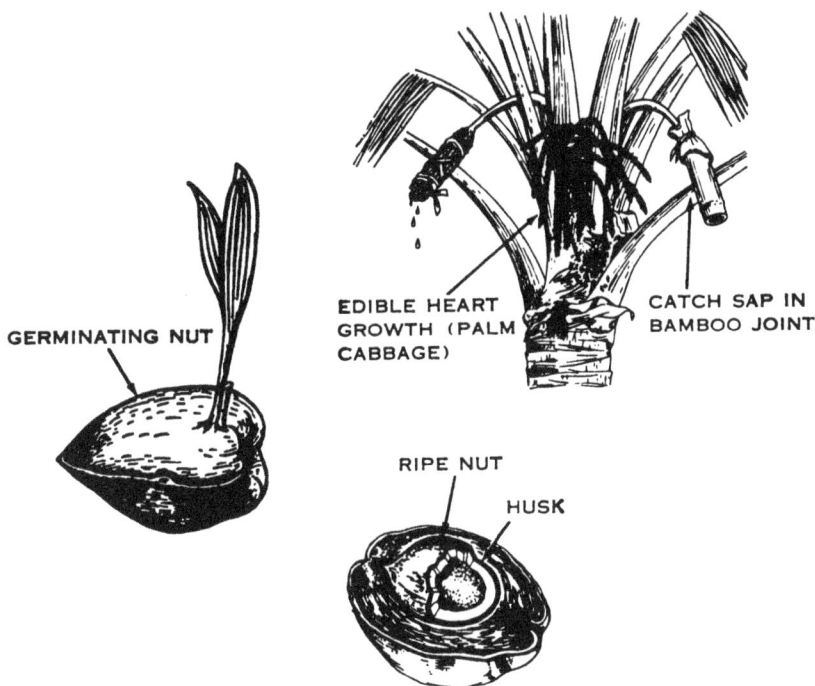

GERMINATING NUT

EDIBLE HEART
GROWTH (PALM
CABBAGE)

CATCH SAP IN
BAMBOO JOINT

RIPE NUT

HUSK

Figure 76. Germinating coconut.

The nuts are hard and dry when mature. Eat them after parching them over coals (fig. 77).

(12) *Cashew nut.* This nut grows in all tropical climates, on a spreading evergreen tree that reaches a height of 40 feet. The leaves are 8 inches long and 4 inches wide; the flowers are yellowish-pink. The fruit is thick, pear-shaped, pulpy, and red or yellow when ripe, with a kidney-shaped nut growing at the tip. This nut incloses one seed and is edible roasted. Be careful of the green hull surrounding the nut. It contains an irritant poison that will blister your eyes and tongue like poison ivy. This poison is destroyed when the nuts are roasted (fig. 78).

e. Seeds and Grains. The seeds of many plants such as buck-wheat, ragweed, amaranth, goosefoot, and the beans and peas from beanlike plants contain oils rich in protein. The grains of all cereals and many other grasses are also rich in plant protein. They may either be ground between stones, mixed with water and cooked to make porridge, or parched. Grains like corn can also be preserved in this manner for future use. Following are some plants with edible seeds and grains:

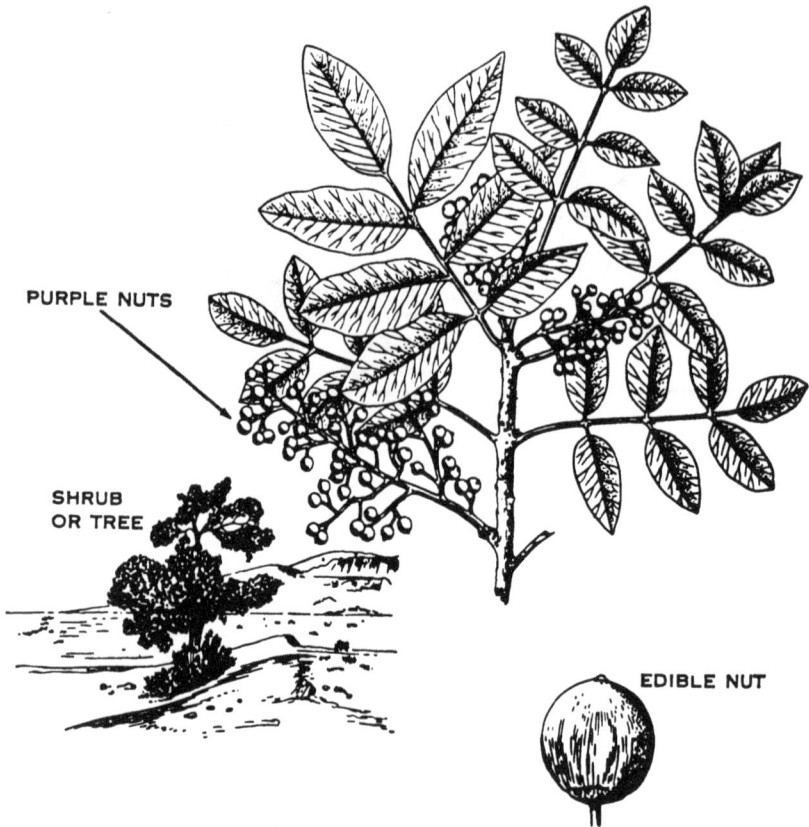

PURPLE NUTS

SHRUB OR TREE

EDIBLE NUT

Figure 77. Pistachio nut.

(1) *Baobab* (*c*(1) above).

(2) *Sorrel* (*c*(6) above.)

(3) *Sea orach.* You will find this plant along seashores from the Mediterranean countries ·to inland areas in North Africa and eastward to Asia Minor and central Siberia. It is thinly branched with small, edible gray-colored leaves about an inch long. The flowers grow in narrow, densely compacted spikes at the tips of the branches (fig. 79).

(4) *St. John's bread.* This tree grows in arid wastelands bordering the Mediterranean Sea on the fringes of the Sahara, across Arabia, Iran, and into India. It is ever-green and reaches a height of 40 to 50 feet. The tree's leaves are leathery and glistening, with 2 to 3 pairs of leaflets, while its flowers are small and red. A seed pod grows on the tree that has a sweet edible pulp. You can pulverize the seeds that are within the pod and cook them as porridge (fig. 80).

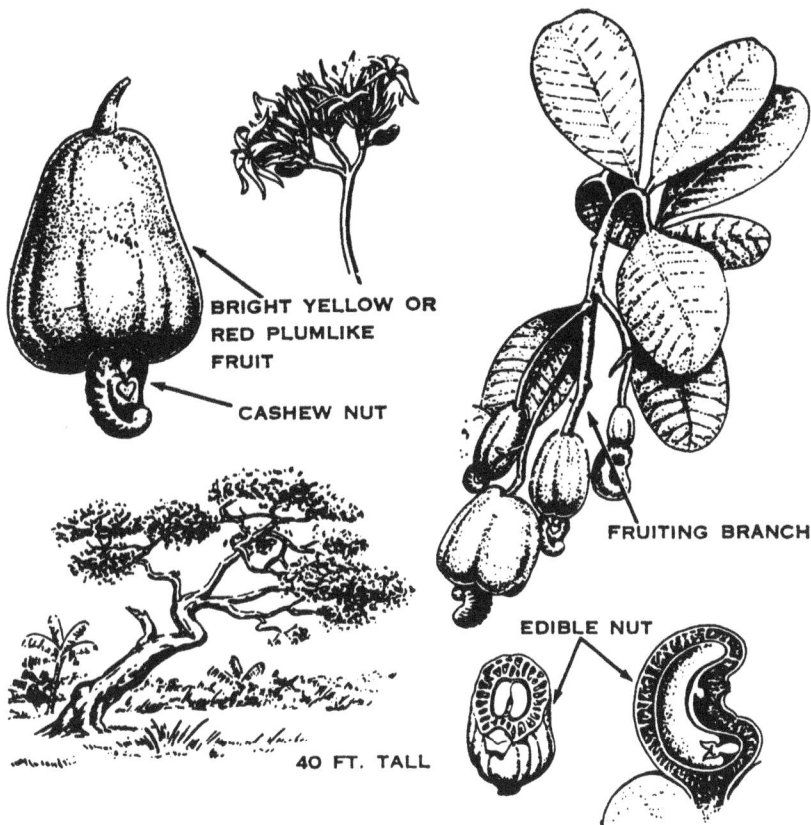

BRIGHT YELLOW OR
RED PLUMLIKE
FRUIT

CASHEW NUT

FRUITING BRANCH

EDIBLE NUT

40 FT. TALL

Figure 78. Cashew nut.

(5) *Luffa* (*b*(2) above).

(6) *Rice.* Rice normally grows in wet areas as a cultivated plant. It is found in tropical and warm, temperate countries throughout the world; however, wild rice exists in Asia, Africa, and parts of the United States. It is a coarse grass growing to a height of 3 to 4 feet with rough hard leaf blades one-half to 2 inches wide. The rice grains grow inside a hairy, straw-colored covering out of which the mature grains shatter when ripe. Roast these rice grains, and beat them into a fine flour. You can carry this as a powder. Combine the flour with palm oil to make cakes. Wrap these in large green leaves and carry them for future use. You can also prepare rice by boiling it (fig. 81).

(7) *Lotus lily* (*c*(9) above).

(8) *Goa bean.*

(a) This plant grows in tropical Africa, Asia, the East

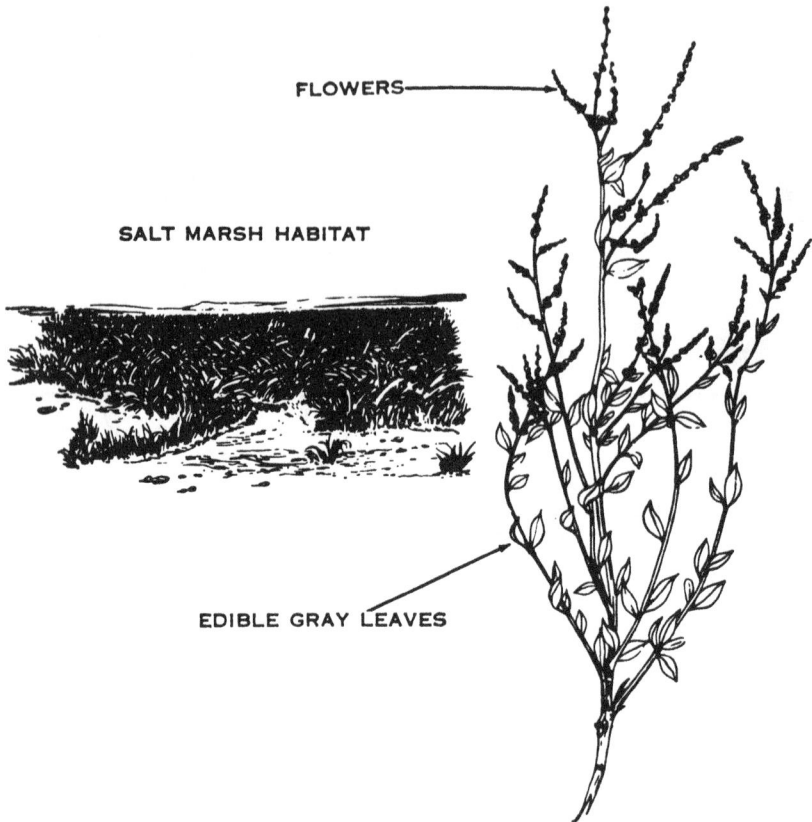

FLOWERS

SALT MARSH HABITAT

EDIBLE GRAY LEAVES

Figure 79. Sea orach.

Indies, the Philippines, and Formosa. It represents an edible bean common to the Old World tropics and is found in clearings and around abandoned gardens (fig. 82).

(b) The goa bean is a climbing plant covering trees and shrubs and has a bean 9 inches long, leaves 6 inches long, and bright-blue flowers. The mature pods are four-angled with jagged wings.

(c) Eat the young pods like string beans and the mature seeds by parching or roasting them over hot coals. Eat the roots raw and the young leaves raw or steamed.

(9) *Bamboo* (b (4) above).

f. *Fruits.*

(1) Edible fruits are plentiful in nature and can be classified as dessert or vegetable. Dessert fruits include the familiar blueberry and crowberry of the North and the

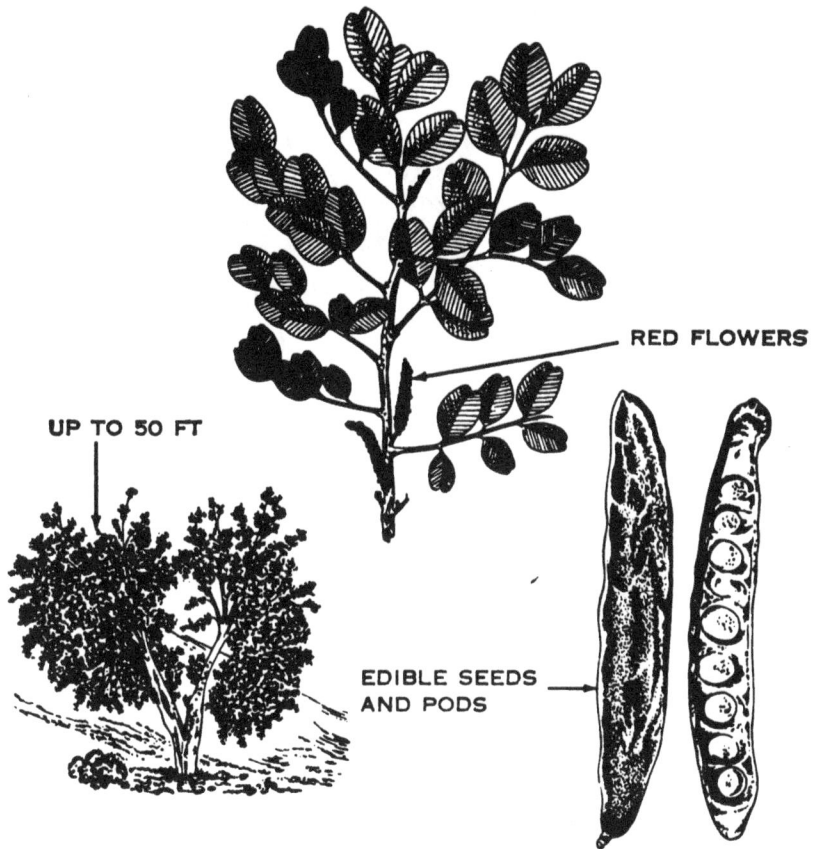

RED FLOWERS

UP TO 50 FT

EDIBLE SEEDS
AND PODS

Figure 80. St. John's bread.

cherry, raspberry, plum, and apple of the temperate
zone. Vegetable fruits are the common cultivated tomato,
cucumber, pepper, eggplant, and okra.

(2) You are probably familiar with many of the wild berries
and fruits of the United States. However, to refresh
your memory on the familiar ones and perhaps increase
your knowledge of those growing in other areas, here
are some that you may contact.

(a) Rose-apple. This tree is native to the Indo-Malayan
region but has been planted widely in most other
tropical countries. This small tree (10 to 30 feet high)
also appears in a semiwild state in thickets, waste
places, and secondary forests. It has tapering leaves
about 8 inches long and greenish-white flowers up to
3 inches across. The fruit is 2 inches in diameter,
greenish or yellow, and has a rose-like odor. It is ex-

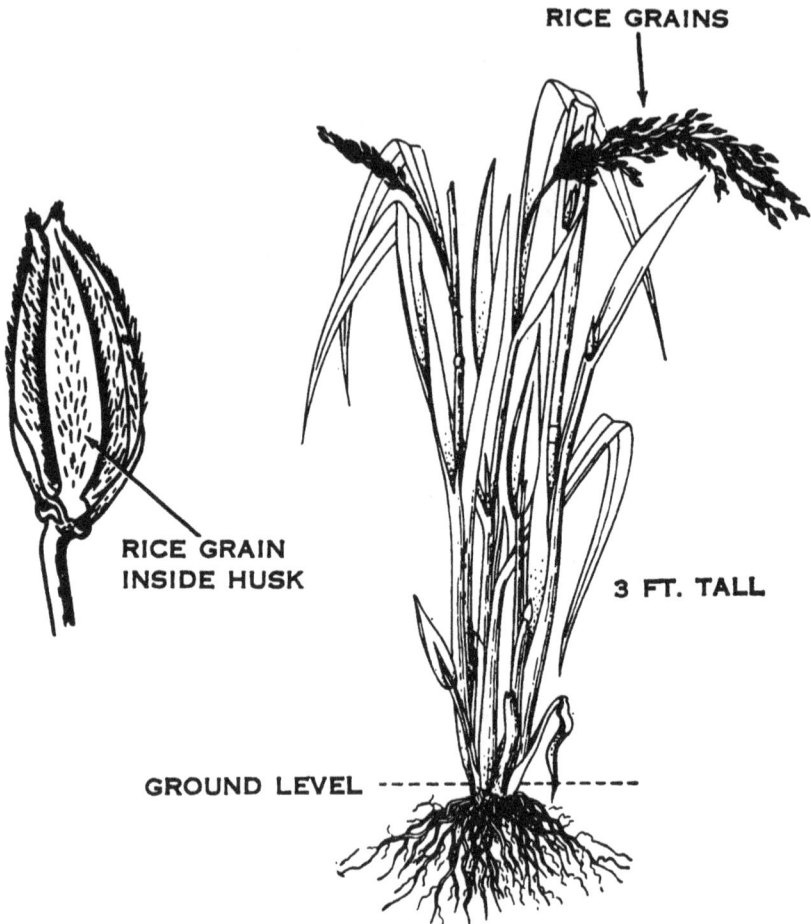

RICE GRAINS

RICE GRAIN INSIDE HUSK

3 FT. TALL

GROUND LEVEL

Figure 81.   Wild rice.

cellent fresh or cooked with honey or palm sap (fig. 83).

(b) Wild huckleberries, blueberries, and whortleberries. Large patches of wild huckleberries overflow with ripe fruit on the tundra in Europe, Asia, and America in late summer. Farther south throughout the northern hemisphere these berries and their close relatives, the blueberry and whortleberry, are common. When they appear in the tundra of the north these wild berries grow on low bushes. Their relatives to the south are borne on taller shrubs which may reach six feet in height. Look for them to be red, blue, or black when ripe (fig. 84).

(c) Cloudberry. The cloudberry is native to the tundra

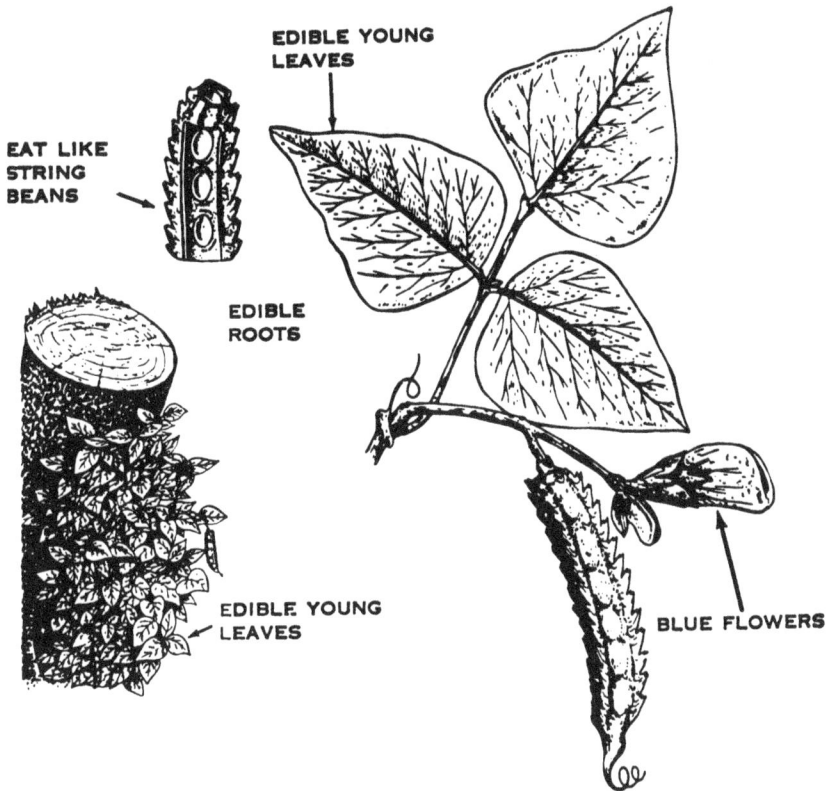

Figure 82. Goa bean.

of Scandinavia, northern Asia, and North America. It grows on an erect plant which covers large areas of ground, and is seldom over one foot high in the southern limit of its distribution, and only a few inches tall on the tundra. The fruit is borne at the top of the plant and is yellow when ripe. It is one of several kinds of wild berries related to the blackberry, raspberry, and dewberry (fig. 85).

(d) Mulberry. Mulberry trees grow in North and South America, Europe, Asia, and Africa. In the wild state they are found in forested areas, along roadsides, and in abandoned fields, and often grow 20 to 60 feet tall. The fruit looks like blackberry and is 1 to 2 inches long. Each berry is about as thick as your finger and varies in color from red to black (fig. 86).

(e) Wild grape vine (fig. 87). This climbing plant is found throughout eastern and southwestern United States, Mexico, Mediterranean areas, Asia, East In-

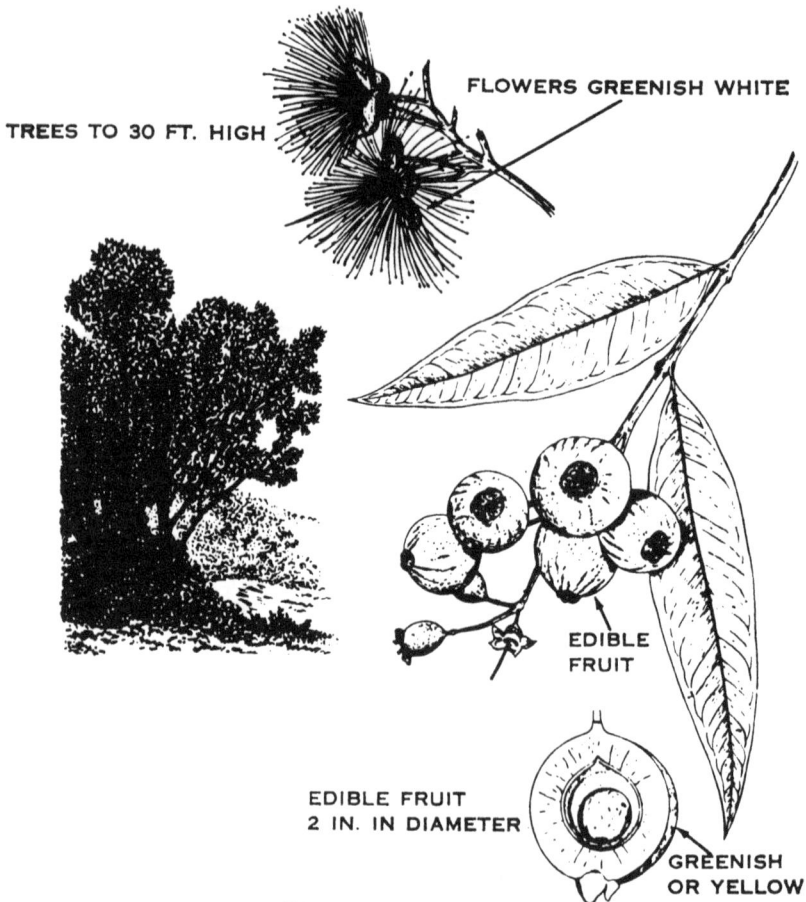

TREES TO 30 FT. HIGH

FLOWERS GREENISH WHITE

EDIBLE FRUIT

EDIBLE FRUIT
2 IN. IN DIAMETER

GREENISH
OR YELLOW

Figure 83. Rose-apple.

dies, Australia, and Africa. It overruns other vegetation where it occurs. Its leaves are deeply lobed and are similar to cultivated grapes. The fruit hangs in bunches and is rich in natural, energy-giving sugar. You can also extract water from the grape vine (ch. 3).

(f) Baobab (c (1) above).

(g) Wild crabapple. This fruit is common in the United States, temperate Asia, and in Europe. Look for it in open woodlands, on the edge of woods, or in fields. The apple looks sufficiently like its tame relative to be easily recognized wherever you might find it. If you wish to store up some food for future use, you can cut this fruit into thin slices an dry it (fig. 88).

(h) Bael fruit. This fruit grows on a small, citrus-type

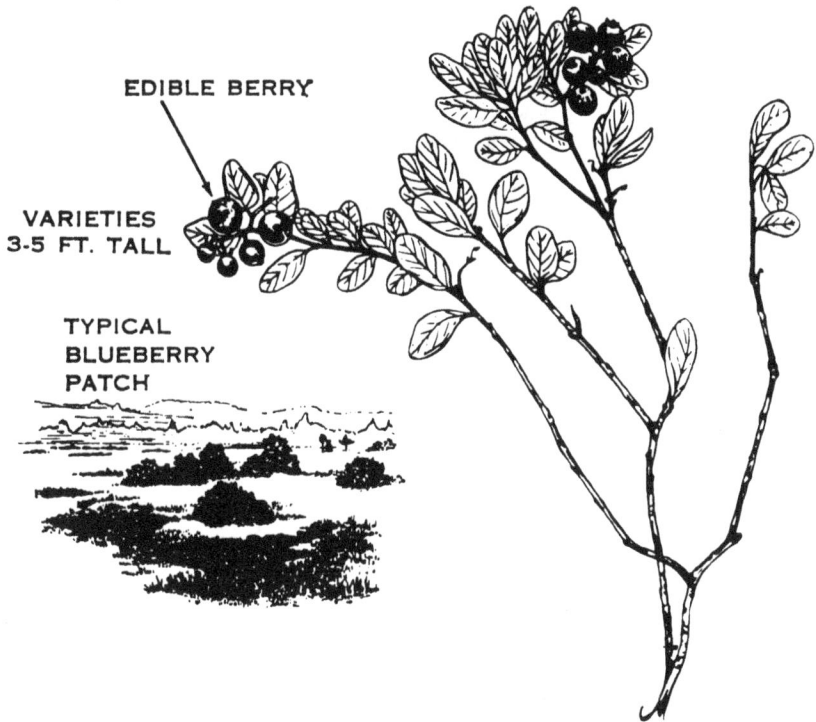

EDIBLE BERRY

VARIETIES
3-5 FT. TALL

TYPICAL
BLUEBERRY
PATCH

Figure 84.   Wild blueberry.

tree and is related to oranges, lemons, and grapefruit. It is found wild in the region of India bordering the Himalayan mountains, in central and southern India, and in Burma. The tree is 8 to 15 feet tall with a dense and spiny growth while the fruit is 2 to 4 inches in diameter, gray or yellowish, and full of seeds. Eat the fruit when it is just turning ripe, or mix the juice with water for a tart but refreshing drink. Like other citrus fruits, this is rich in vitamin C (fig. 89).

(i) Wild fig. Most of the 800 varieties of wild figs grow in tropical and subtropical areas having abundant rainfall; however, a few desert kinds exist in America. The trees are evergreen with large, leathery leaves. Look in abandoned gardens, along roadways and trails, and in fields for a tree with long aerial roots growing from its trunk and branches. After identifying the tree, look for the fig fruit which grows out directly from the branches. The fruit resembles a child's top or a pear. Many varieties are hard and

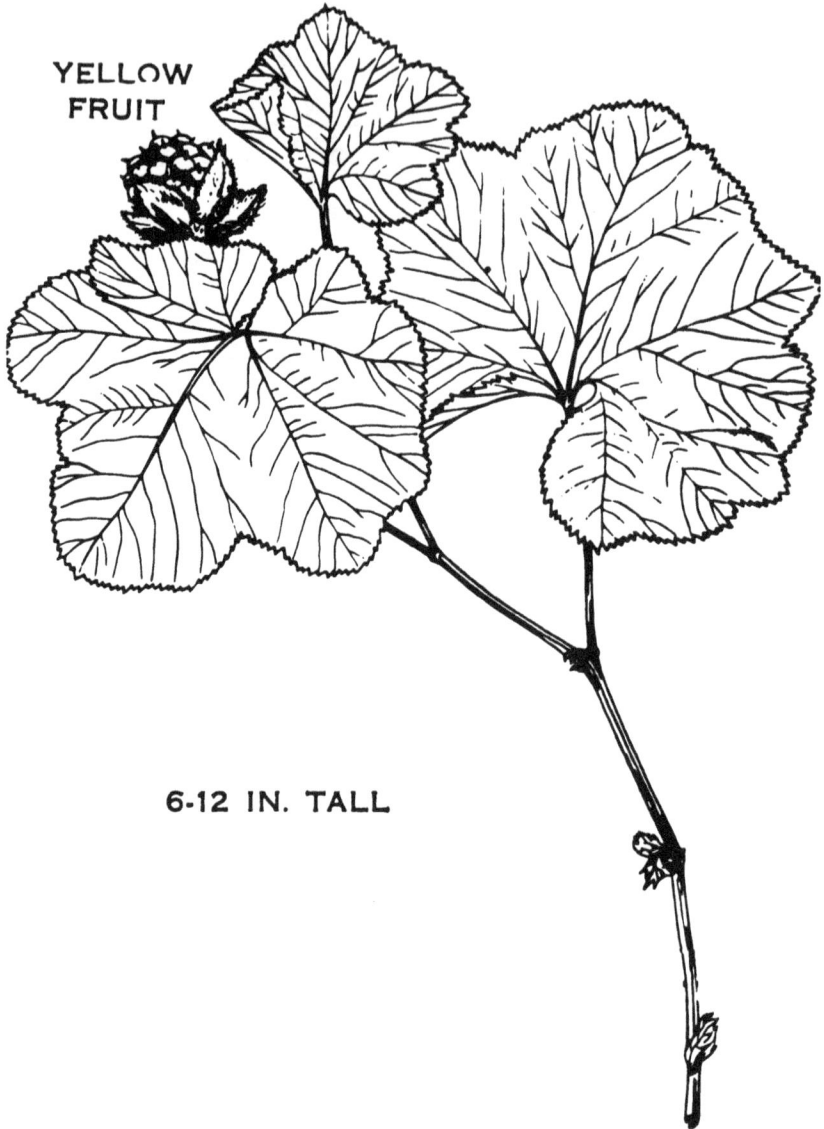

YELLOW
FRUIT

6-12 IN. TALL

Figure 85. Cloudberry.

woody and covered with irritating hairs, and otherwise worthless as a survival food. The edible type is
soft when ripe, almost hairless, green, red, or black
in color (fig. 90).

(3) Plants with vegetable-type fruit include—

    (a) Wild caper. This plant grows either as a spring
shrub or small tree about 20 feet tall in North Africa,
Arabia, India, and Indonesia. It is leafless with spine-

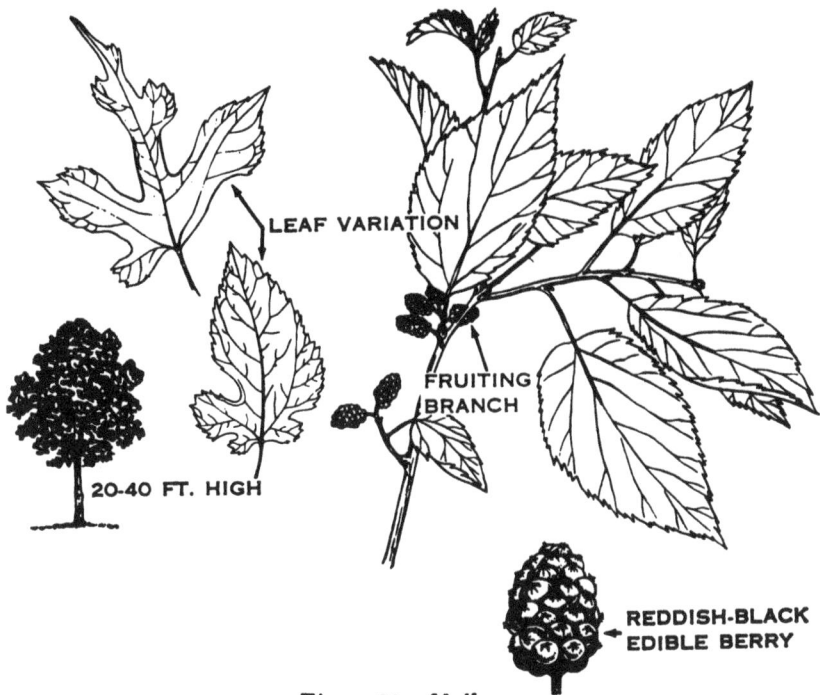

LEAF VARIATION

20-40 FT. HIGH

FRUITING BRANCH

REDDISH-BLACK EDIBLE BERRY

Figure 86.   Mulberry.

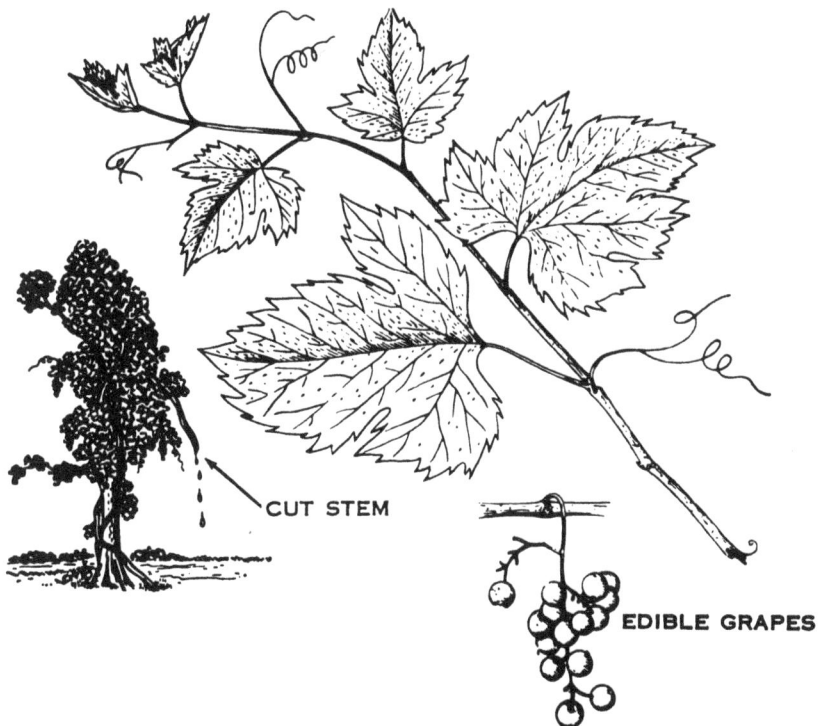

CUT STEM

EDIBLE GRAPES

Figure 87.   Wild grape vine.

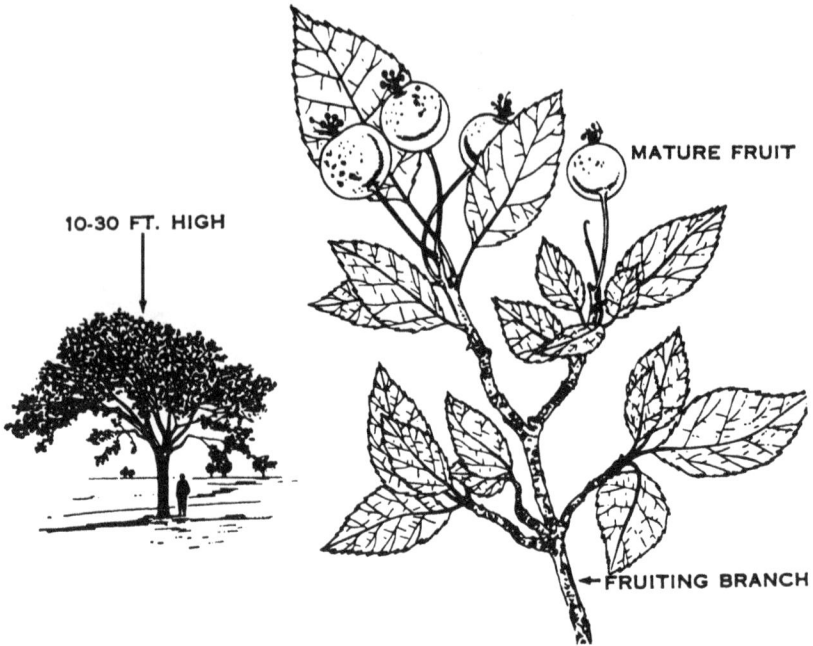

10-30 FT. HIGH

MATURE FRUIT

←FRUITING BRANCH

Figure 88.   Wild crabapple.

covered branches, flowers and fruit that grow near
the tips of the branches. Eat the fruit as well as the
flower buds (fig. 91).

(b) Breadfruit. The breadfruit is a common tropical
tree. It grows up to 40 feet tall with leathery leaves
1 to 3 feet long (fig. 92). The fruit is delicious when
ripe and you can prepare it using the following
methods: Eat the fruit raw or grilled on the embers of
an open fire. Scrape it lightly with your knife or to
eat raw, remove the skin first; then pick off the lumps
of flesh with your fingers, separating the seeds, and
discarding the hard outer covering. To prepare for
grilling, scrape the fruit and remove the stalk.

(c) Wild gourd (*b*(2) above).

(d) Water plantain (*a*(2)*(c)* above.)

g. *Barks.*

(1) The inner bark of a tree—the layer next to the wood—
may be eaten raw or cooked. You can even make flour
from the inner bark of cottonwood, aspen, birch, willow,
and pine tree by pulverizing it. Avoid the outer bark
because of the presence of large amounts of tannin.

(2) Pine bark is rich in vitamin C. Scrape away the outer

GREENISH WHITE FLOWERS

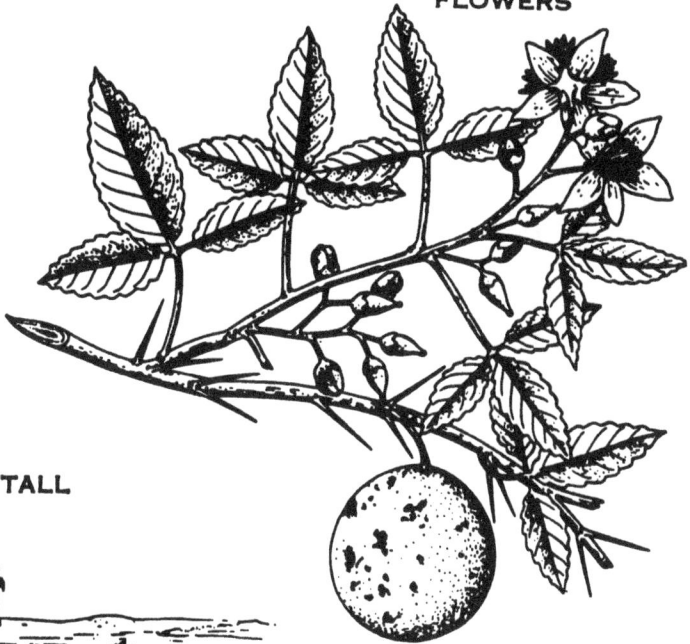

8-15 FT. TALL

GRAY OR YELLOWISH EDIBLE FRUIT, 2-4 IN. IN DIAMETER

Figure 89. Bael fruit.

bark and strip the inner bark from the trunk. Eat it fresh, dried, or cooked, or pulverize it into flour.

h. *Fungi.*

(1) About 16,000 varieties of edible fungi grow in different parts of the world. The mushrooms you eat on your steaks and the mold in the blue cheese that you spread on crackers are two forms of fungi.

(2) Although fungi are not a good substitute for meat, they are comparable in food content to common leafy vegetables, and they often are available in areas where other edible plants are scarce.

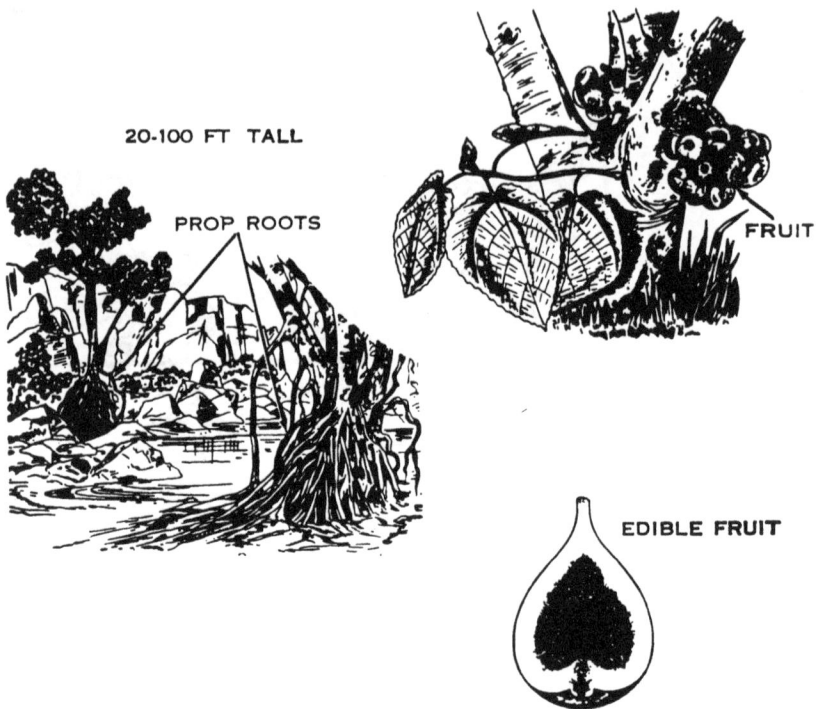

20-100 FT TALL

PROP ROOTS

FRUIT

EDIBLE FRUIT

Figure 90. Wild fig.

(3) Gilled fungi, or mushrooms, are the most common of
edible fungi and are 98 percent safe to eat, but they
have been subjected to many "old wives' tales" and, thus
are considered untouchable foods by many people. The
term "toadstools," for example, has been used so much
to describe any inedible or poisonous variety of mushroom that people apply this name to most unfamiliar
varieties. The distinguishing characteristics of "toadstools" such as odor, peeling of skin bruises, and livid
colors may also be present in the edible forms. The
best way to tell the difference is to study the general
characteristics of both the edible and poisonous varieties. Supplement this information with the following
list of hints for selecting edible mushrooms:

(a) Dig the gilled mushroom completely out of the ground.
Eliminate those having a cup, or vulva, at the base
(fig. 93).

(b) Avoid all gilled mushrooms in the young stage. This
young mushroom will have a button-like appearance
and may be distinguished from the young edible puff-

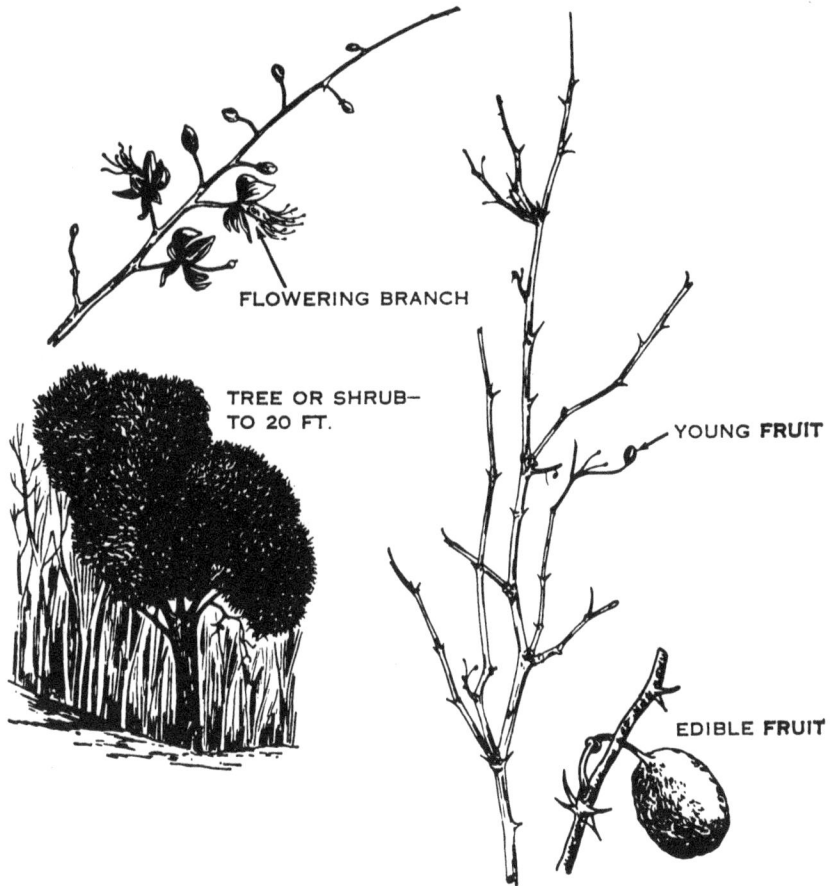

Figure 91. Wild caper.

ball because it will have a short stem, not a characteristic of the puffball.

(c) Avoid all ground-growing mushrooms with the under-
side of the cup full of minute reddish spores (fig. 94).

(d) Avoid gilled mushrooms with membrane-like cups or
scaly bulbs at the base, especially if the gills are
white.

(e) Avoid all gilled mushrooms with white or pale milky
juice.

(f) Avoid all gilled woodland mushrooms with a smooth,
flat, reddish top and white gills radiating out from
the stem-like spokes.

(g) Avoid yellow or yellowish-orange mushrooms growing on old stumps. If they have crowded and solid
stems, convex overlapping cups, broad gills extend-

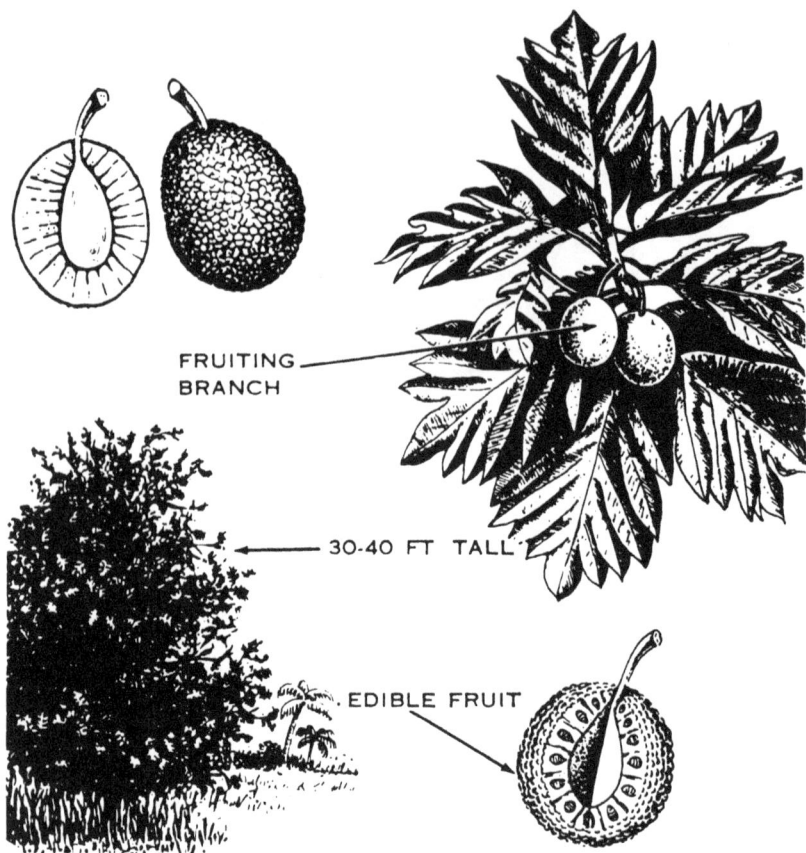

Figure 92.   Breadfruit.

ing irregularly down the stem, or surfaces that glow
phosphorescently in the dark, they are probably poisonous.

(h) Avoid any mushroom which seems to be too ripe,
water-soaked, spoiled, or maggoty.

(i) Become familiar with the following poisonous mushrooms of the Amanite family (the most deadly is the
Death Angel). See figures 95 and 96.   The Death
Angel is widespread in Europe, Asia, and America
but seems to be more common in north temperate
regions (figs. 97 and 98). This plant produces one
of the tox-albumin poisons, the poisons found in rattlesnakes and other venomous animals and the poisons
that produce death in cholera and diphtheria. The
amount of this fungus necessary to produce death is
small.

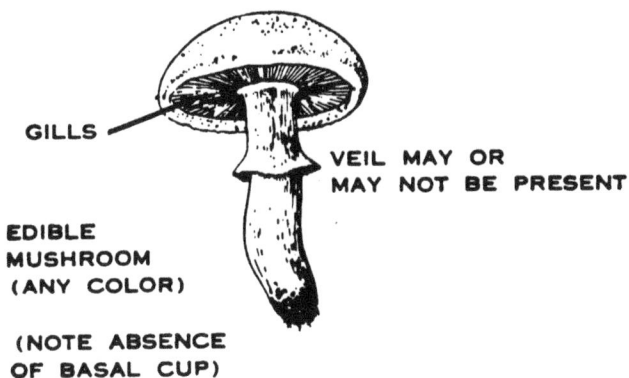

GILLS

VEIL MAY OR
MAY NOT BE PRESENT

EDIBLE
MUSHROOM
(ANY COLOR)

(NOTE ABSENCE
OF BASAL CUP)

Figure 93. Mushroom with gills, veil, but no basal cup.

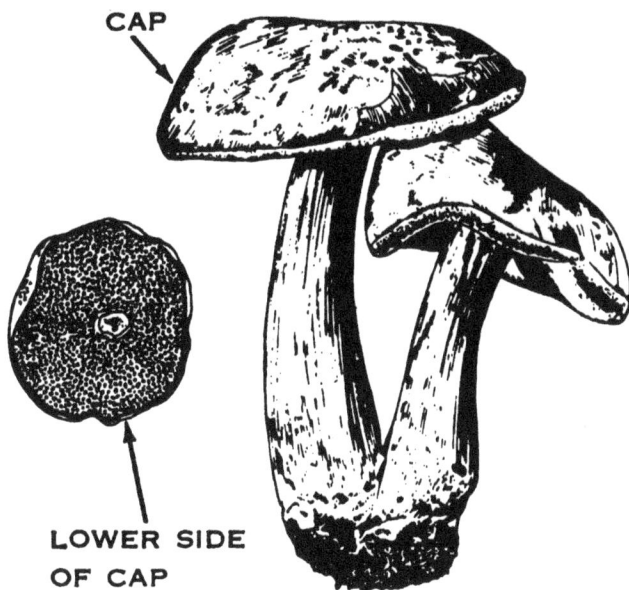

CAP

LOWER SIDE
OF CAP

Figure 94. Boletus cap and the area in which to look for reddish spores.

(j) If you get sick after eating mushrooms, tickle the back of your throat to induce vomiting. Do not drink water until after you vomit; then drink lukewarm water and powdered charcoal.

(4) All nongilled fungi are nonpoisonous when eaten fresh. A familiar example of a nongilled fungus is the puffball. Others include morels, coral fungi, coral hydnums, and cup fungi. These are illustrated below in figures 99 through 103.

REDISH-WHITE CAP WITH WHITE FLECKS

Fly agaric

Figure 95. Fly agaric.

i. Seaweeds.

(1) Properly prepared seaweed found near or on the shores of the larger ocean areas is quite edible and can be an important part of your diet. It is a valuable source of iodine and vitamin C.

(2) When you select seaweed for food, choose plants attached to rocks or floating free because those that have lain on the beach for any length of time may be spoiled or decayed. You can dry the thin, tender varieties over a fire or in the sun until they are crisp; and then crush and use them for soup flavoring. Wash the thick leathery seaweeds, and soften them by boiling. Eat these varities with other foods.

(3) Following are some edible seaweeds that you might find:

(a) Green seaweeds, often called sea lettuce, grow in the

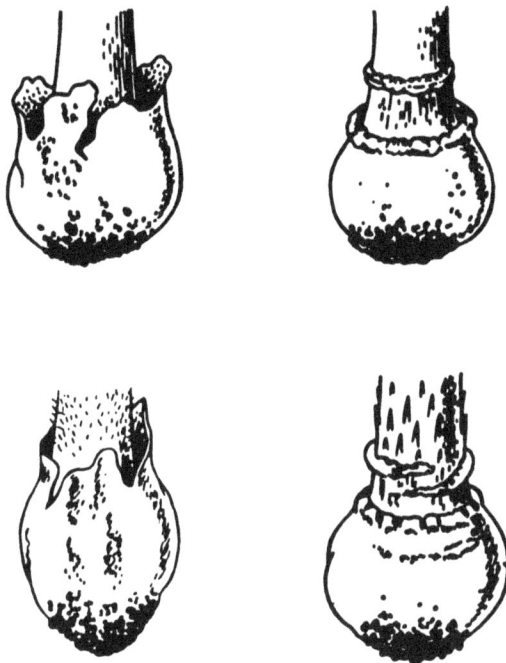

Figure 96. *Identifying cups of fly agaric.*

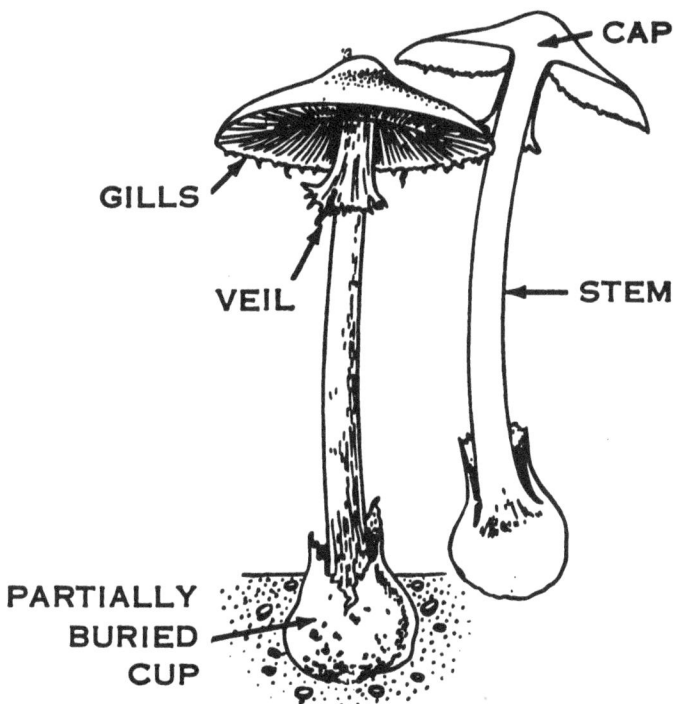

Figure 97. *Death Angel with gills, veil, stem, and cup.*

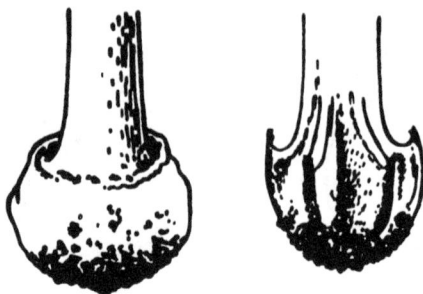

Figure 98. Identifying cups of Death Angel.

(CHALK-WHITE INSIDE)

(1-12 IN. IN DIAMETER)

Figure 99. Puffball.

(ASHEN-GRAY)

Figure 100. Morel (Ashengray).

Pacific and North Atlantic Oceans. Wash in clean water and use them just as you would garden lettuce (fig. 104).

(b) Edible brown seaweeds include—

  1. Sugar wrack. The young stalks of this plant are sweet and are found on both sides of the Atlantic and on the coasts of China and Japan (fig.105).

  2. Kelp. This seaweed is found in both the Atlantic

(WHITE, ORANGE, YELLOW,
PALE VIOLET, BUFF)
(2-6 IN. HIGH)

Figure 101. Coral fungus (2 to 6 inches high, white, orange, yellow, pale violet, or buff).

(FOUND ON DEAD WOOD)

(WAXY WHITE)

Figure 102. Coral Hydnum (dead wood).

(1-3 IN. HIGH)

Figure 103. Cup fungus.

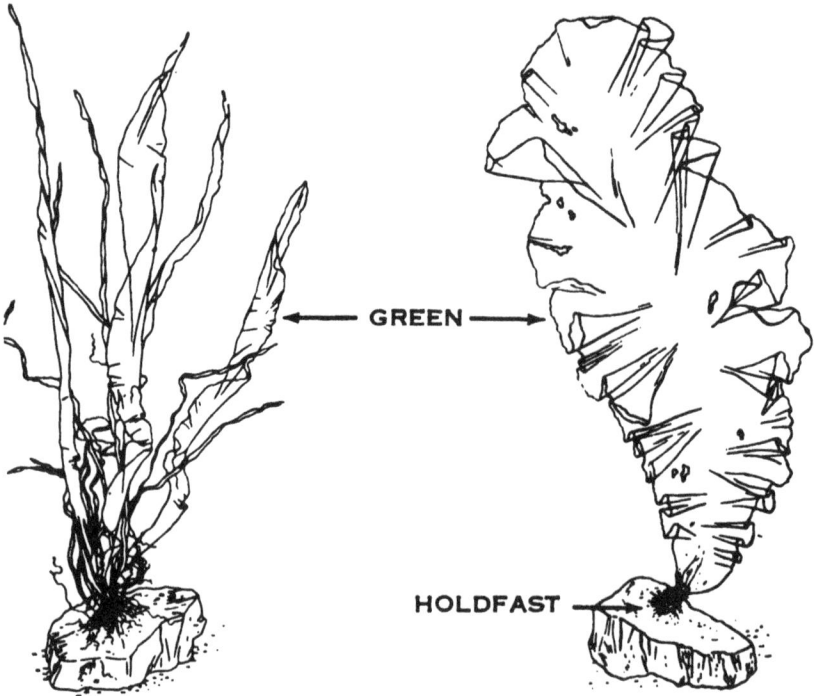

GREEN

HOLDFAST

Figure 104. Sea lettuce.

and Pacific Oceans below the high tide line on submerged ledges and rocky bottoms. It has a short
cylindrical stem and thin, wavy, olive-green or
brown fronds from one to several feet long. Boil it
before eating; then mix with vegetables or soup
(fig. 106).

3. Irish moss. This moss is found on both sides of the
Atlantic. It is tough, elastic, and leathery and you
can find it below the high tide line or upon the
shore. Boil before eating (fig. 107).

STALKS 1-5 FT. LONG

OLIVE GREEN
OR BROWN

HOLDFAST

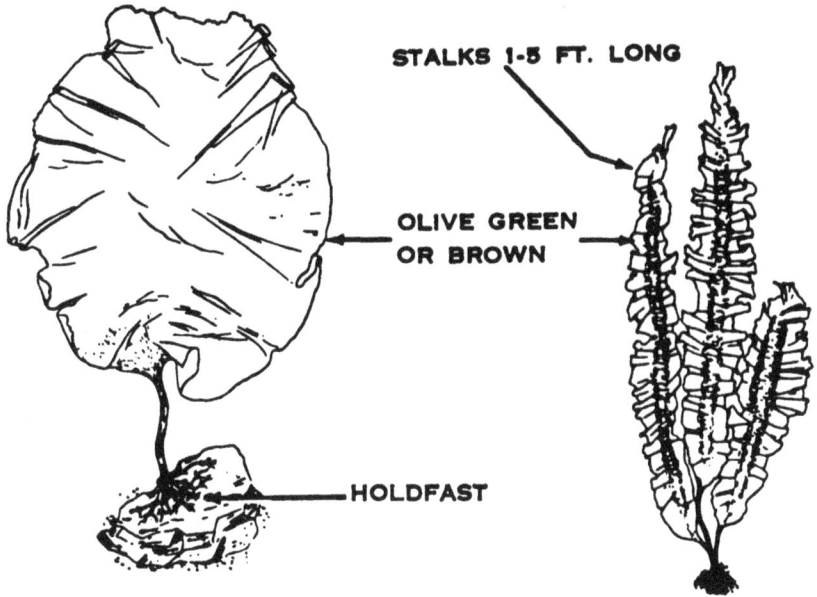

Figure 105. Sugar wrack.

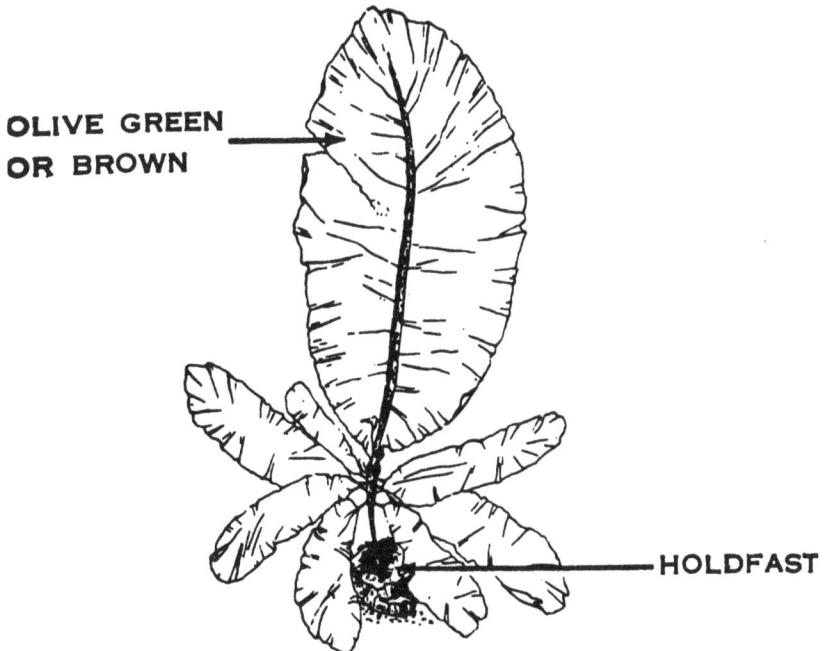

OLIVE GREEN
OR BROWN

HOLDFAST

Figure 106. Kelp.

REDDISH TO WHITE

HOLDFAST ATTACHES TO ROCK

Figure 107. Irish moss.

(c) Red seaweeds have a characteristic reddish tint and include—

1. Dulse. This has a short stem which quickly broadens into a thin, broad, fan-shaped leaf. The leaf is dark red and divided by several clefts into short, round-tipped lobes. Varying from a few inches to a foot in length, this plant is attached to rocks or coarser seaweeds and is found on the Atlantic and Mediterranean coasts. It is sweet to taste and can be dried and rolled and used as chewing tobacco (fig. 108).

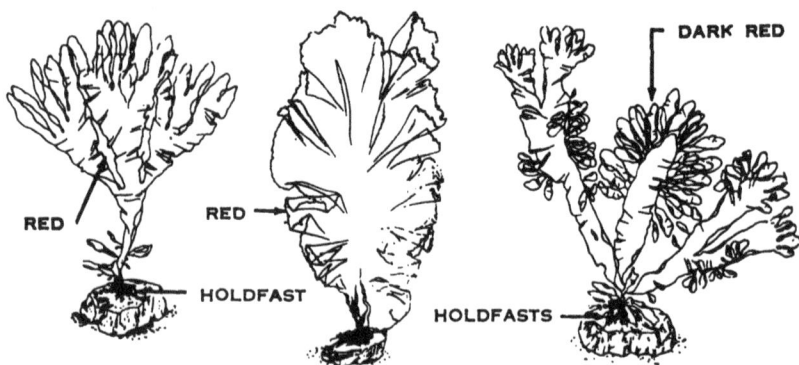

DARK RED

RED

RED

HOLDFAST

HOLDFASTS

THREE FORMS OF DULSE

Figure 108. Dulse.

2. Laver. This seaweed is common to the Atlantic and Pacific areas and is usually red, dark purple, or purplish-brown with a satiny sheen or filmy luster. Use it as a relish, boil it gently until tender, or

pulverize it; add it to crushed grains and fry it in the form of flatcakes. Look for this plant on the beach at low tide (fig. 109).

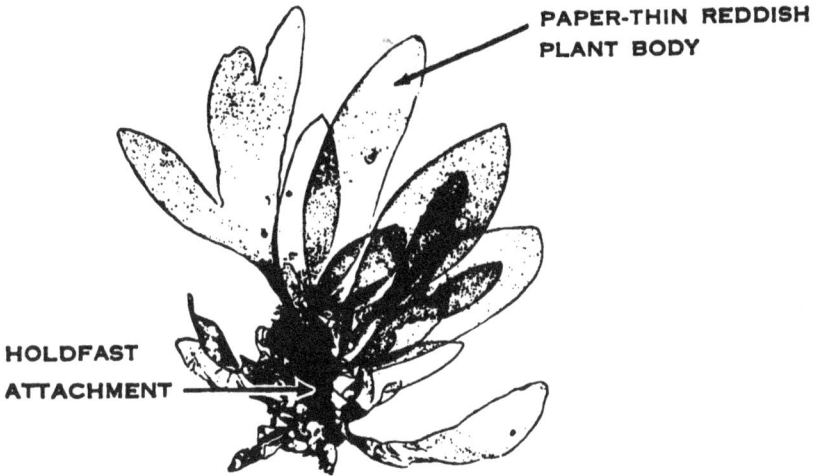

PAPER-THIN REDDISH PLANT BODY

HOLDFAST ATTACHMENT

Figure 109.   Laver.

(d) Freshwater algae is a variety of seaweed common to China, America, and Europe. One of the more familiar varieties is nostoc and you will find it during the spring in pools. It forms green, round, jelly-like lobules about the size of marbles. Dry this plant and use it in soups (fig. 110).

GELATINOUS

NOSTOC GROWS ON GROUND

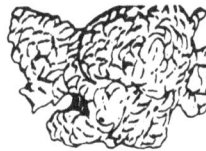

BLUE-GREEN COLONIES SPREAD OVER GROUND IN GRASSY MEADOWS

Figure 110.   Freshwater algae.

35. Cultivated Vegetables

Harvested vegetable or grain fields are rich sources of food. Keep your eyes open for old potato, corn, and turnip fields and harvested pea patches in Europe and the temperate countries of Asia.

a. If you discover a potato field, dig into the hills for potatoes that were overlooked when the field was harvested. Eat the potatoes raw or cooked.

b. Look for fields with vegetable stalks that remain in the ground after harvest. These include turnips, rutabagas, carrots, beets, and radishes. You can eat any of these cooked or raw However, peel or otherwise clean these foods before eating to eliminate the dangers caused by contamination from fertilizers.

c. In abandoned corn fields, search the ground for discarded ears. Eat kernels raw, cooked, or parched as pinole, a highly nutritious food made by parching corn in hot ashes or over a fire and grinding it into flour. A handful of this in water makes a nutritious and tasty combination.

Section III. ANIMAL FOODS

36. Varieties

a. Many people consider grasshoppers, hairless caterpillars, wood-boring beetle larvae and pupae, spider bodies, and termites as delicacies. You have probably eaten some of them unknowingly in your daily food. The time may come when you have no choice but to eat insects such as these. If so, you will find them much more palatable if you cook them until they are dry or disguise them in a stew.

b. Foods derived from animals have much more food value per pound than those derived from plants, but are more difficult to obtain. A knowledge of edible animals, including where to look for them and how to catch them, increases your survival chances.

37. Foods From Fresh Water

Fresh water lakes, ponds, streams, and rivers are abundant food reservoirs. Look for them wherever you are; they support more animal life in a smaller area than the land, and often the food they harbor is easier to acquire. You can count on finding such water animals as fish, frogs, snails, and crabs around or in most inland waters.

a. Fish. Of the animal life around or in fresh water, fish are probably the most difficult to catch; so don't expect too much at one sitting. It may take hours or even days before you are successful. It can be done though, even with crude equipment, if you are patient, and know where, when, and how to fish.

    (1) *When to fish.* It is difficult to state the best time to fish because different species feed at different times, both day and night. As a general rule, look for fish to feed just before dawn and just after dusk; just before a

storm as the front is moving in; and at night when the moon is full or waning. Rising fish and jumping minnows may also be signs of feeding fish.

(2) *Where to fish.* The place you select to start fishing depends on the type of water available and the time of day. In fast running streams in the heat of the day, try deep pools that lie below the riffles. Toward evening or in the early morning, float your bait over the riffle, aiming for submerged logs, undercut banks, and overhanging bushes. On lakes in the heat of the summer, fish deep as fish seek the coolness of deeper water. In the evening or early morning in summer, fish the edges of the lake; fish are more apt to feed in shallow water. Lake fishing in the spring and late fall is more productive on the edge in shallow water because fish are either bedding or seeking warmer water. With practice you can locate the beds of some species of fish by their strong, distinctively "fishy" odor.

(3) *Bait.*

    (a) As a general rule, fish bite bait taken from their native water. Look in the water near the shores for crabs, fish eggs, and minnows and on the banks for worms and insects. If you hook a fish, inspect its stomach to see what it has been eating; then try to duplicate this food. Use its intestines and eyes for bait if other sources are unproductive. If you use worms, cover the hook completely. With minnows, pass the hook through the body of the fish under its backbone in the rear of the dorsal fin. Be sure you do not sever the minnow's backbone.

    (b) You can make artificial bait from pieces of brightly colored cloth, feathers, or bits of bright metal fashioned to duplicate insects, worms and minnows. Strive to make your artificial bait look natural by moving it slowly or copying the actions of natural fish food.

(4) *Making hooks and line.* If you have no hooks, improvise them out of insignia, pins, bone, or hardwood (fig. 111). By twisting bark or cloth fibers you can fashion a sturdy line. Using the inner bark of a tree, knot the ends of two strands and secure them to a solid base. Hold a strand in each hand and twist clockwise, crossing one above the other counterclockwise. Add fiber as necessary to increase the length of the line. If you have parachute shroud lines available, use these for your line.

(5) *Catching fish.* There will be times when the most

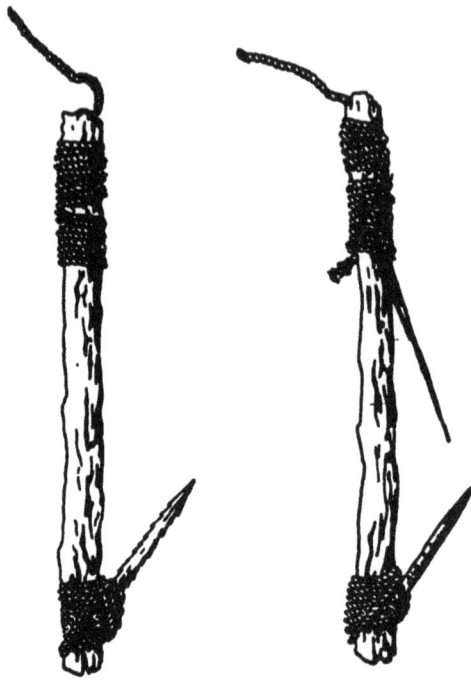

BONE   **NAIL**

Figure 111. Improvised hooks and lines.

elaborate line and suitable bait will not yield a single fish. Do not become discouraged because there are other methods that may prove more productive.

(a) Set lines.

1. Set lines provide a practical method for catching fish if you happen to be "holed-up" for a period of time near a lake or stream awaiting an opportunity to continue your trip safely. Simply tie several hooks onto your line. Bait them and fasten the line to a low-hanging branch that will bend when a fish is hooked. Keep this line in the water as long as you are in the area, checking it periodically to remove fish and rebait the hooks.

2. An excellent hook for a set line is the gorge or skewer hook (fig. 112). Sink the skewer into a chunk of bait. After the fish swallows the bait, the skewer swings crosswise and lodges in the stomach, securing the fish to the line.

(b) Jigging. This method requires an 8- to 10-foot limber

BAITED SKEWER

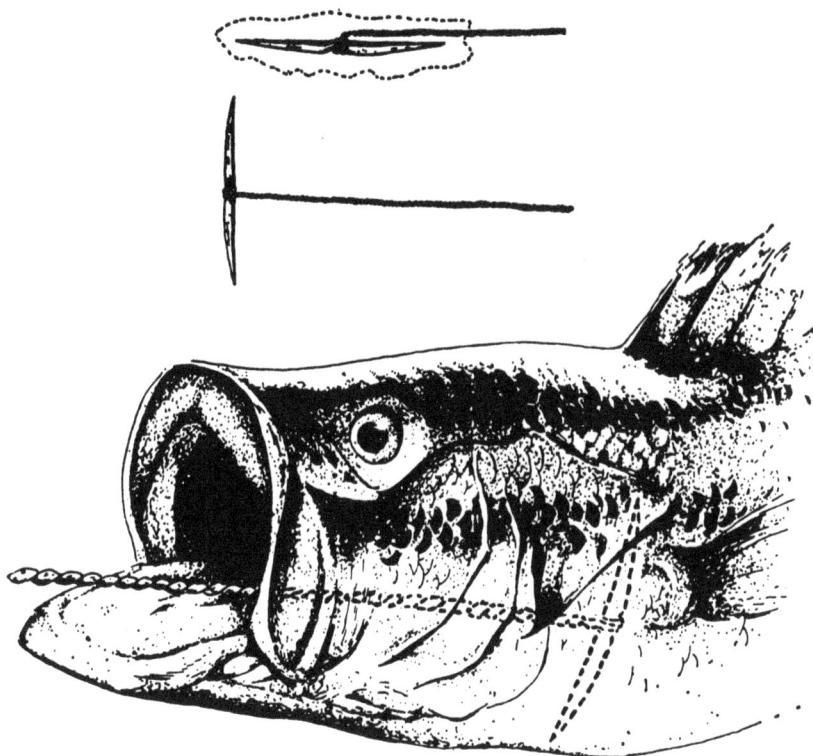

Figure 112. Skewer hook.

cane or similar type pole, a hook, a piece of brilliant
metal shaped like a commercial fishing spoon, a 2- to
3-inch strip of white meat or pork rind or fish intes-
tine, and a piece of line about 10 inches long. Attach
the hook just below the spoon on the end of the short
line, and tie the line to the end of the pole. Working
close to the edge near lily pads or weed beds, dabble
the hook and spoon apparatus just below the surface
of the water. Occasionally slap the water with the
tip of the pole to attract large fish to your bait. This
method is especially effective at night.

(c) *Using your hand.* This method is effective in small
streams with undercut banks or in shallow ponds left
by receding flood waters. Place your hands in the
water and allow them to reach water temperature.
Reach under the bank slowly, keeping your hands
close to the bottom if possible. Move your fingers

slightly until you contact a fish. Then work your hand gently along its belly until you reach its gills. Grasp the fish firmly just behind its gills.

(d) *Muddying.* Small isolated pools caused by the receding waters of flooded streams are often abundant in fish. Disturb the mud of the bottoms of these puddles by stamping in them or using a stick until the fish are forced to seek clearer water at the surface. Then throw them out with your hands or club them.

(e) *Spearing.* This method is difficult except when the stream is small and the fish are large and numerous; during spawning season; or when the fish congregate in pools. Tie your bayonet on the end of a pole; sharpen a piece of bamboo; lash two long thorns on a stick; or fashion a bone spear point, and position yourself on a rock over a fish run. Wait patiently and quietly for a fish to swim by. Select only the large fish.

(f) *Nets.* The edges and tributaries of lakes and streams are usually abundant with fish too small to hook or spear but large enough to net. Select a forked sapling and make a circular frame. Stitch or tie your undershirt, or tie the cloth-like material found at the base of coconut trees to this frame, making sure the bottom is closed. Scoop upstream around rocks or in pools with this improvised net.

(g) *Traps or weir.*

1. These are useful for catching both fresh and salt water fish, especially those that move in schools. In lakes or large streams, fish approach the banks and shallows in the morning and evening. Sea fish traveling in large schools regularly approach the shore with the incoming tide, often moving parallel to the shore and guiding on obstructions in the water.

2. A fish trap (figs. 113–116) is an inclosure with a blind opening where two fence-like walls extend out like a funnel from the entrance. The time and effort you put into building a trap depends upon your need for food and the length of time you plan to stay in one spot.

3. If you are near the sea, pick your trap location at high tide and build it at low tide. On rocky shores use natural rock pools. On coral islands use natural pools on the surface of reefs by blocking the open-

ings as the tide recedes. On sandy shores, use sand-bars and the ditches they inclose. Fish in the lee of offshore sandbars. Build your trap as a low stone wall extending out into the water and forming an angle with the shore.

4. In small, shallow streams, make your fish trap with stakes or brush set into the stream bottom so that the stream is blocked except for a small narrow opening into a stone or brush pen. Wade in and herd the fish into your trap. Catch or club them when they get into shallow water.

Figure 113. Tidal fish trap.

(h) *Shooting.*

1. If you are fortunate and have a weapon and sufficient ammunition try shooting fish. Aim slightly under the fish in water that is less than three feet deep.

2. A hand grenade exploded in a school of fish will supply you with food for days. Dry or otherwise preserve those that you do not eat fresh (ch. 5).

(i) *Poisoning.* Throughout the warm regions of the world there are various plants and other materials which natives use for poisoning fish. The active poison in these is harmful only to cold blooded animals. Fish poisons include—

1. *The derris plant.* This woody vine grows in South-

127

Figure 114. Maze-type fish trap.

east Asia. Powder the roots and throw them into the stream at its head waters if possible. In a short time the stunned fish will rise to the surface (fig. 117).

2. *The barringtonia tree.* This tree is found near the sea shore in Malaya and parts of Polynesia. Crush the seeds and throw them into the stream or pond (fig. 118).

3. *Coral and sea shells.* Lime will kill fish. Burn coral and sea shells together to obtain this fish poison.

(j) *Ice fishing.* You can obtain fish in the winter by fishing through a hole in the ice. Keep the hole open by covering it with brush and heaping loose snow over the cover.

1. Fish tend to gather in deep pools, so cut your ice holes over the deepest part of the lake. Place a rig

Figure 115. V-type trap.

similar to the one in figure 119 at several holes. When the flag moves to an upright position remove the fish and rebait the rig.

2. Take a 3-foot pole and a string long enough to reach the bottom of the place where you fish. Make a small spoon shaped spinner from a C-ration can or from any piece of bright metal. Attach an improvised fish hook to the line and tie the improvised spinner to the line just above the hook. When fishing move the rod in an up and down motion in such a way that the bright metal object vibrates. Fish close to the place where the shelf near the shore drops off to lake bottom, at the edge of the reeds, or close to some projecting rock formation.

3. You might be able to see fish swimming beneath clear ice. Strike the ice sharply with a rock or log above the fish. This will stun them long enough for you to cut a hole and remove the fish.

b. Frogs, Newts, and Salamanders. These small amphibian

*Figure 116. Arrow shape
fish trap.*

animals inhabit areas surrounding fresh water in warm and
temperate climates throughout the world.

    (1) Hunt frogs at night when you can locate them by their
croaking. Club them or snag the larger ones on a hook
and line. Eat the entire body after skinning (fig. 120).
See chapter 5.

    (2) Newts and salamanders are found under rotten logs or
under rocks in areas where frogs are abundant.

  c. Mollusks.

    (1) These include fresh and salt water invertebrates such
as snails, clams, mussels, bivalves, periwinkles, chitons,
and sea urchins (fig. 121). See chapter 6 for a dis-

Figure 117.   The derris plant.

cussion of mollusks. Most members of this group are edible; however, be sure that you have a fresh mollusk and that you boil it. If you eat one raw you are inviting parasites into your body.

(2) In fresh water look for these food sources in the shallows, especially in water with a sandy or mud bottom. Near the sea, wait for low tide and check in tidal pools or in the sand.

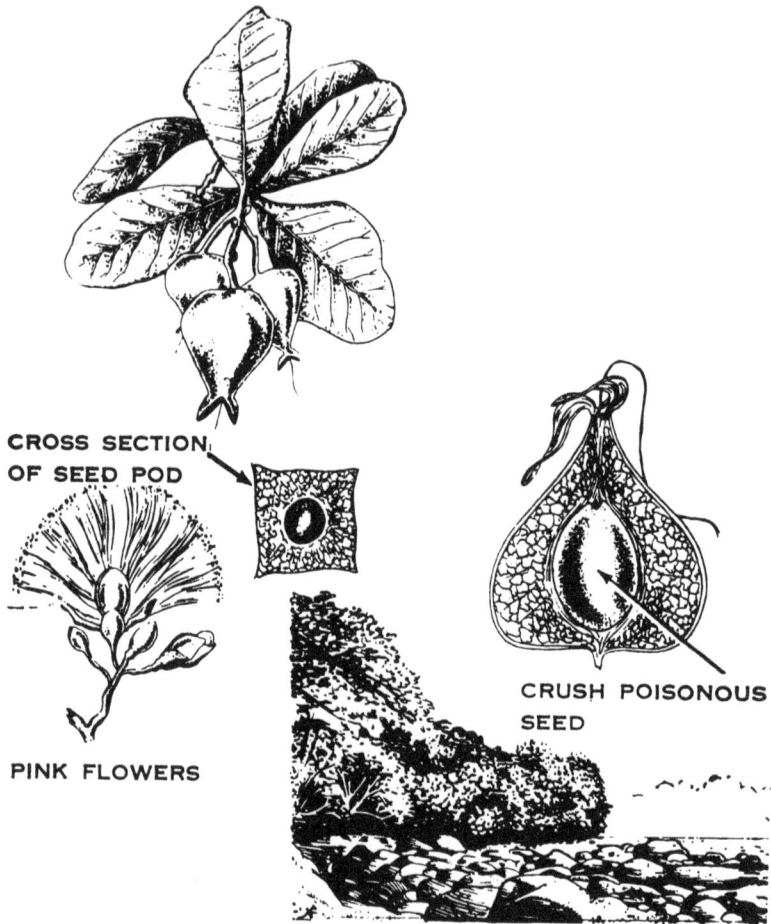

CROSS SECTION
OF SEED POD

CRUSH POISONOUS
SEED

PINK FLOWERS

Figure 118. Barringtonia tree.

d. Crustaceans. Fresh and salt water crabs, crayfish, lobsters, shrimps, and prawns are included in this class. Most of them are edible, but they spoil rapidly and some harbor harmful parasites. Look for them in moss beds under rocks or net them from tidal pools (fig. 123). Fresh water shrimp are abundant in tropical streams, especially where the water is sluggish. Here they cling to branches or vegetation. Cook the fresh water forms; eat the salt water varieties raw if you desire (1, 2, 3, 4, fig. 122).

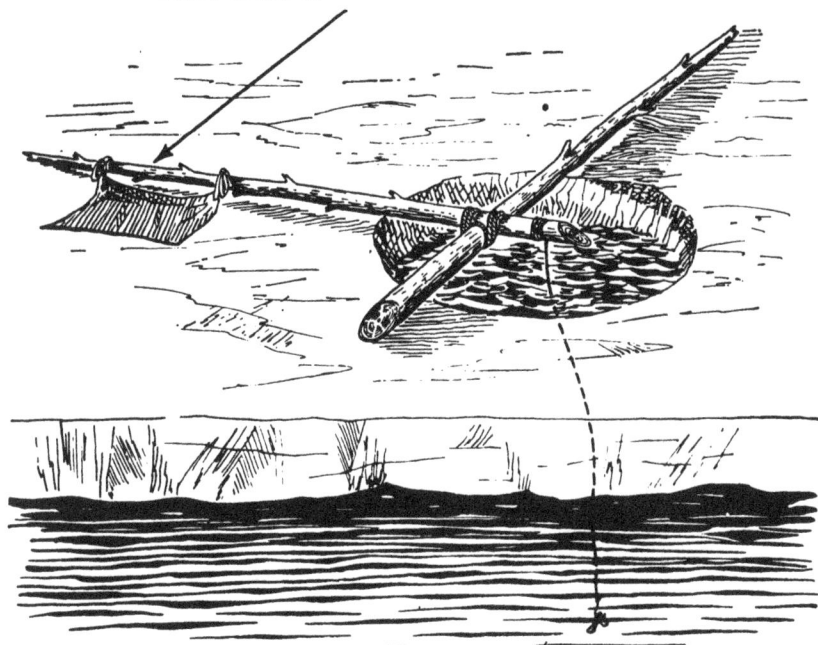

WHEN FISH TAKES BAIT
FLAG GOES UP

Figure 119. Automatic fisherman.

Figure 120. Frog.

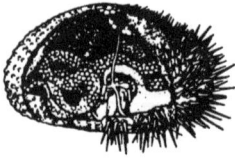

SEA URCHIN

SEA CUCUMBER

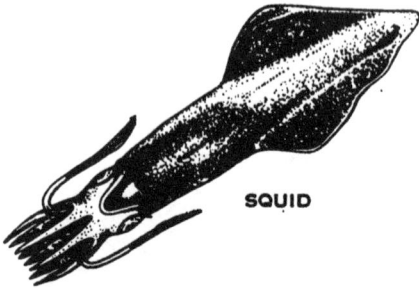

SQUID

OCTOPUS

Figure 121.   Invertebrates.

MUSSELS

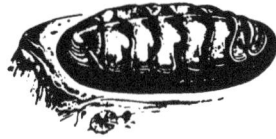

CHITON

SNAIL

OYSTERS

LIMPETS

RAZOR CLAM

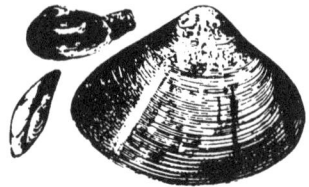

CLAMS

Figure 121—Continued.

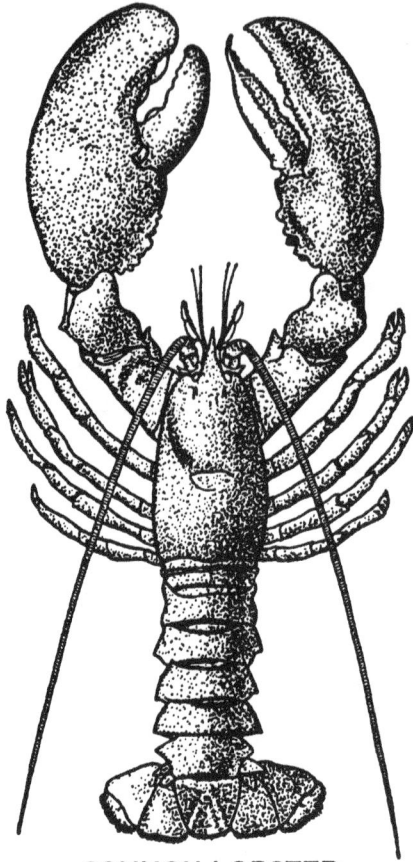

COMMON LOBSTER

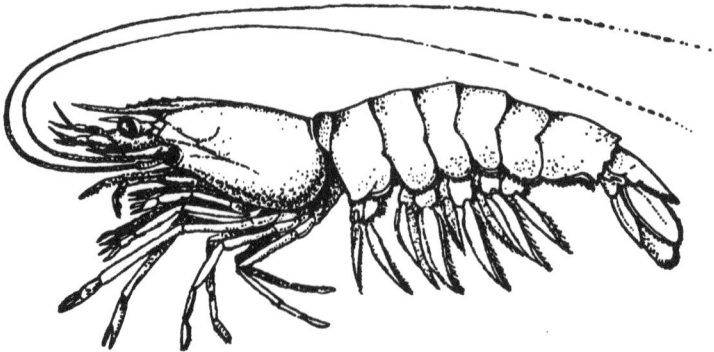

SALT WATER SHRIMP

①

Figure 122.   Crustaceans.

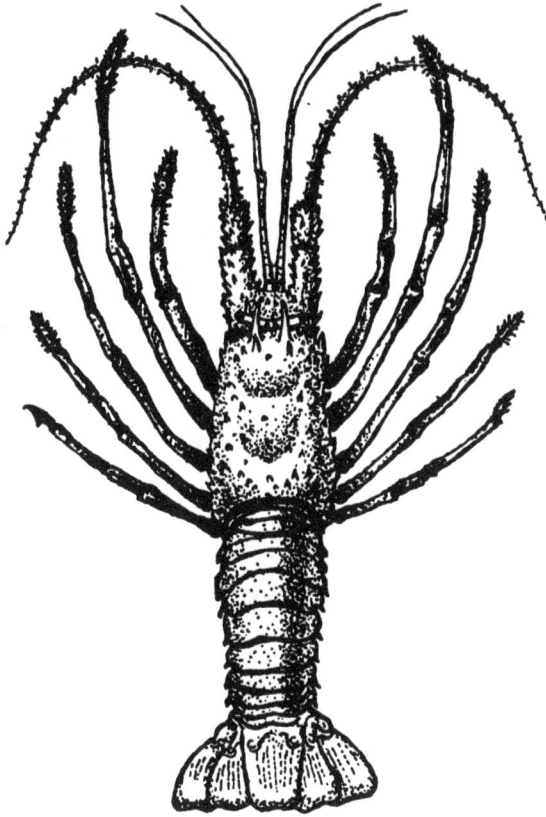

SPINY LOBSTER

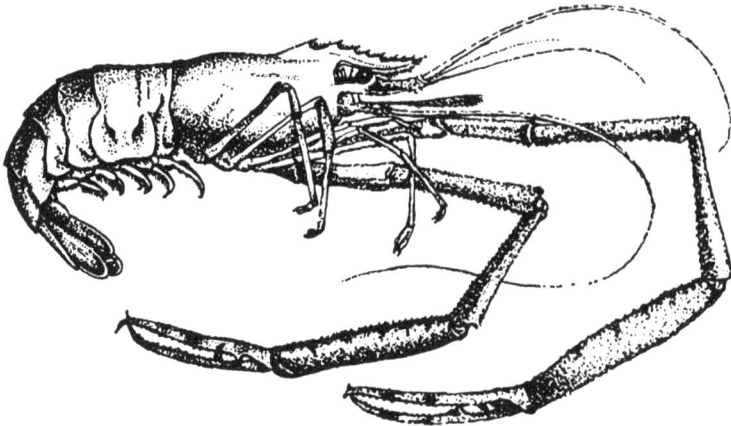

FRESH WATER SHRIMP

②

Figure 122—Continued.

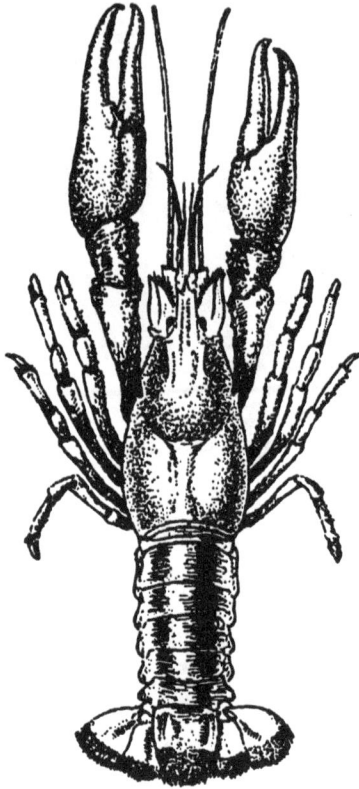

CRAYFISH

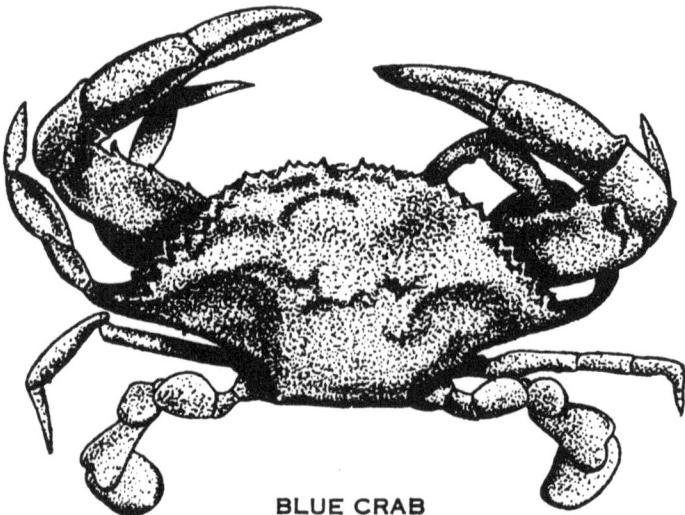

BLUE CRAB

Figure 122—Continued.

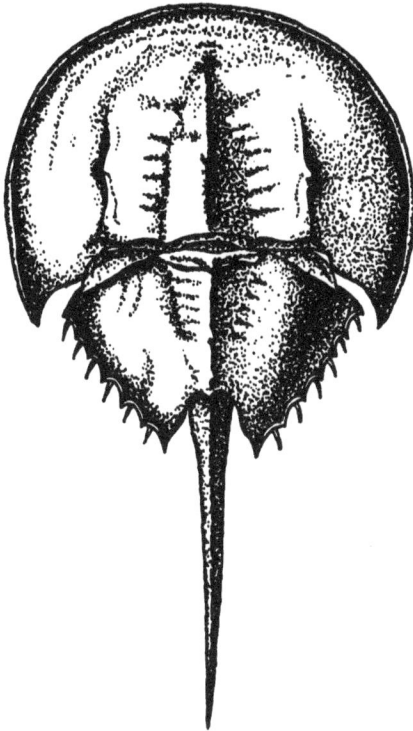

HORSESHOE CRAB

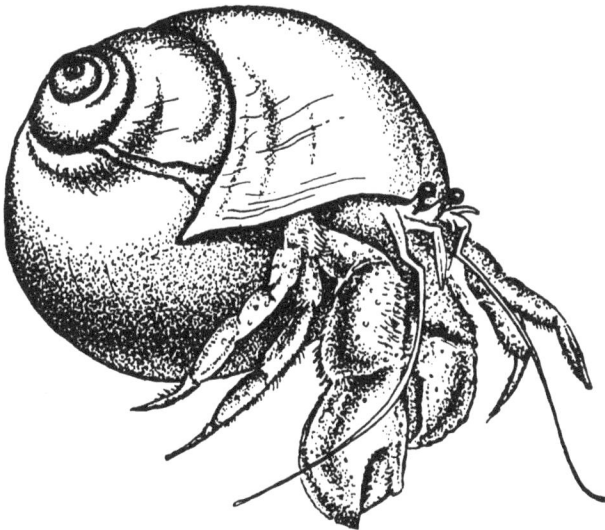

HERMIT CRAB

④

Figure 122—Continued.

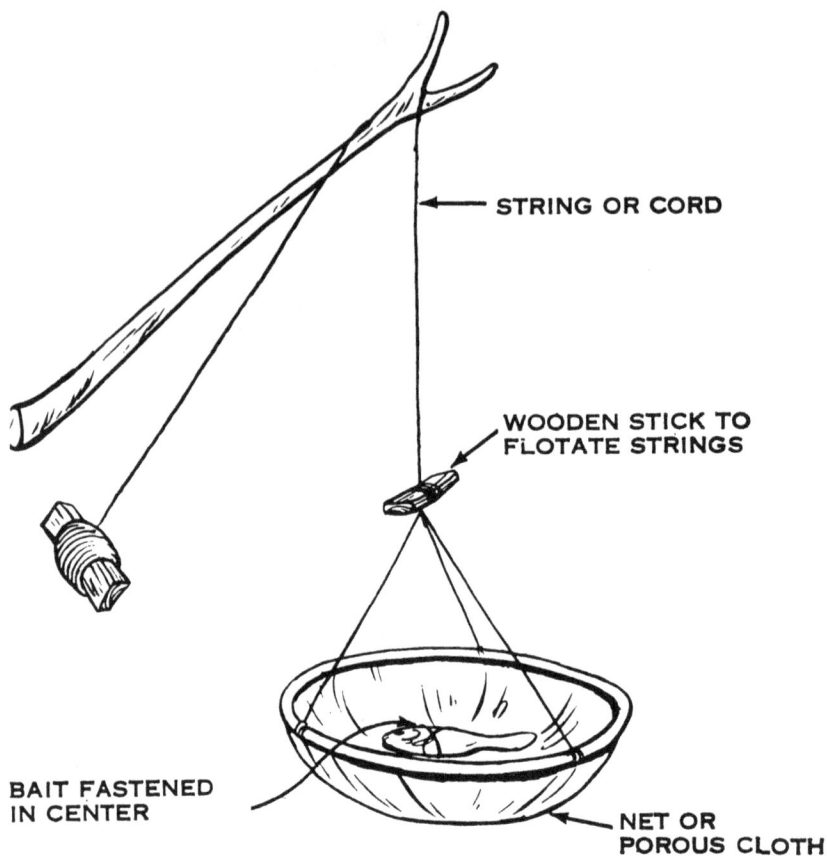

STRING OR CORD

WOODEN STICK TO FLOTATE STRINGS

BAIT FASTENED IN CENTER

NET OR POROUS CLOTH

Figure 123. Improvised crayfish trap.

38. Reptiles

Do not discount snakes, lizards, alligators, and turtles as possible food sources (fig. 124). See chapter 7.

a. Fresh water snakes, both poisonous and nonpoisonous, frequent lakes and streams where the water is sluggish and the banks are covered with driftwood and overhanging branches. Although snakes are edible, use extreme caution when you search for them, especially in areas having poisonous varieties (ch. 7).

b. Lizards are inhabitants of the tropics and subtropics. Included in this group are the two poisonous lizards mentioned in chapter 7, and alligators. All are edible. Remove the sealy skin and then boil or fry the meat; heat alligators over a fire before skinning to loosen the plates.

c. Marine fresh water and land turtles are edible (ch. 6) and are found on land or in waters of the temperate and tropical zones. Club the smaller fresh water turtles, or catch them on a

Figure 124. Alligator.

line. Be careful with the larger snapping ones because they can inflict a serious bite.

39. Insects

Grubs, grasshoppers, termites, and most other insects have food value and are palatable if prepared properly. Use them to provide stock for soup or to add protein to stews (fig. 125). Be sure to cook grasshoppers to kill parasites contained in their bodies.

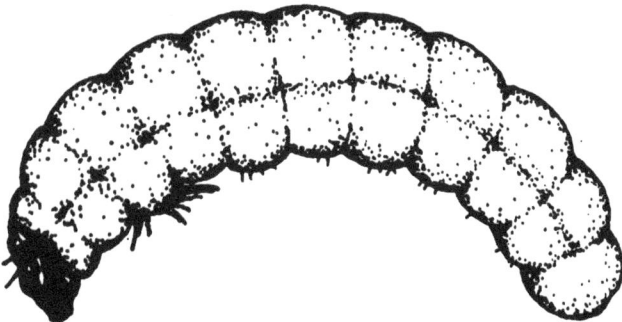

Figure 125. Grub.

40. Birds and Mammals

a. *General.*

(1) All mammals and birds are edible but are probably the most difficult to obtain of all survival food. Consequently, concentrate on the sources mentioned above while considering methods of catching bigger game.

(2) Hunting animals and birds is not an easy job for even the most experienced woodsman; therefore, as a beginner, "still hunt." Find a place where animals pass—a trail, watering place, or feeding ground. Hide nearby, always downwind so the animal can't smell you, and wait for game to come to within range of your weapon or to walk into your trap. Remain absolutely motionless.

(a) If you decide to stalk an animal, do so upwind, moving slowly and noiselessly only when he is feeding or looking the other way. *Freeze* when he looks your way.

(b) Hunt in the early morning or at dusk and look for animal signs such as tracks, a game run, trampled underbrush, and droppings. Remember—animals depend upon their keen sense of sight, hearing, and smell to warn them of danger.

(3) Birds can see and hear exceptionally well but are lacking in their sense of smell. During nesting periods they are less fearful of man. Because of this you can catch them easier in the spring and summer, especially in temperate or arctic areas. They nest in cliffs, branches, marshes, or trees and by watching the older birds you can locate the young or eggs.

b. *Hunting* (ch. 6).

(1) *Finding game.* The secret to hunting successfully is seeing your quarry before he sees you, so keep alert. Watch for signs that tell of the presence of game. As you approach a ridge, lake, or clearing slow down and peer first at distant then closer ground. At water holes that show signs of game, hide and wait until an animal approaches even though it takes hours. In general, apply the military principles of movement and concealment that you have learned.

(2) *Shooting game.* If you have a weapon and see a chance to use it, whistle sharply to encourage your quarry to stop, giving you a chance for a standing shot. On large game aim for a neck, lung, or head shot. In the event you wound an animal and it runs, follow its blood trail

slowly but deliberately. If the quarry is wounded severely, it will lie down soon if not followed; and when it lies down, it will usually weaken and be unable to rise. Approach it slowly and finish it off. After killing a large animal such as deer, gut and bleed it immediately. Cut the musk glands from between its hind legs and at the joints of its hind legs. Be careful not to burst the bladder while removing it.

c. *Trapping.*

    (1) *Know your game.*

        (a) Before you can trap with any luck, you must decide what you wish to trap; what the animal will do; and then catch him doing it. Determine the kind of food he eats and bait your trap accordingly.

        (b) Rats, mice, rabbits, and squirrels are easy to trap. These small mammals have regular habits and confine themselves to limited areas of activity. Simply locate a hole or run and bait and set your trap ((2) – (5) below).

    (2) *Trapping hints.* Following are some tricks that may increase your "take" if you decide to trap game or birds:

        (a) To catch a mammal that lives in hollow trees, try inserting a short forked stick in the hole and twisting so that his loose skin will wrap around the fork. Keep the stick taut while pulling it out.

        (b) Smoke burrow-living animals out of their dens; then using a noose attached to the end of a long pole, snare the quarry as it emerges from the hole.

        (c) Use the noose method to snare birds that are sitting on eggs or roosting. After finding a roosting or nesting area, conceal yourself and wait quietly for the bird to return. Slip the noose quickly over the bird's head and pull to the rear and upward.

        (d) Bait a fish hook with a minnow and place it on the shore near the water. Chances are a bird will snatch it.

        (e) Set snares or traps at night in runways containing fresh tracks or droppings. If you have used a spot for butchering an animal, set a snare in the area. Use animal entrails for bait.

        (f) If you are still without food after experimenting with these methods, set the woods or grassland on fire and wait for the game to break through. Do not use this method except as a last resort.

    (3) *Hanging snare.* Fasten a slip noose to the end of a

bent sapling. Open the noose wide enough to fit over the animal's head but not wide enough for its body to slip through. Secure the trigger so that it holds the sapling as shown in figure 126. Make it sufficiently loose so that a slight jerk on the noose will free the trigger.

(4) *Simple drag noose.* This simple snare is basic to successful survival. It is effective for catching small game and birds.

(5) *Fixed snare.* This snare is particularly useful for catching rabbits. Fasten the loop to a log, tree, or forked stake and set it near a bush or limb as shown in figure

Figure 126. Hanging snares.[6]

[6]Aviation Training, Office of the Chief of Naval Operations, U. S. Navy, "How To Survive On Land and Sea," Copyright 1943, 1951 by The United States Naval Institute.

127. After the trap is sprung the animal will strangle itself.

(6) *Treadle spring snare.* This is effective for small mammals and birds. Cover the treadle with leaves or grass (fig. 128).

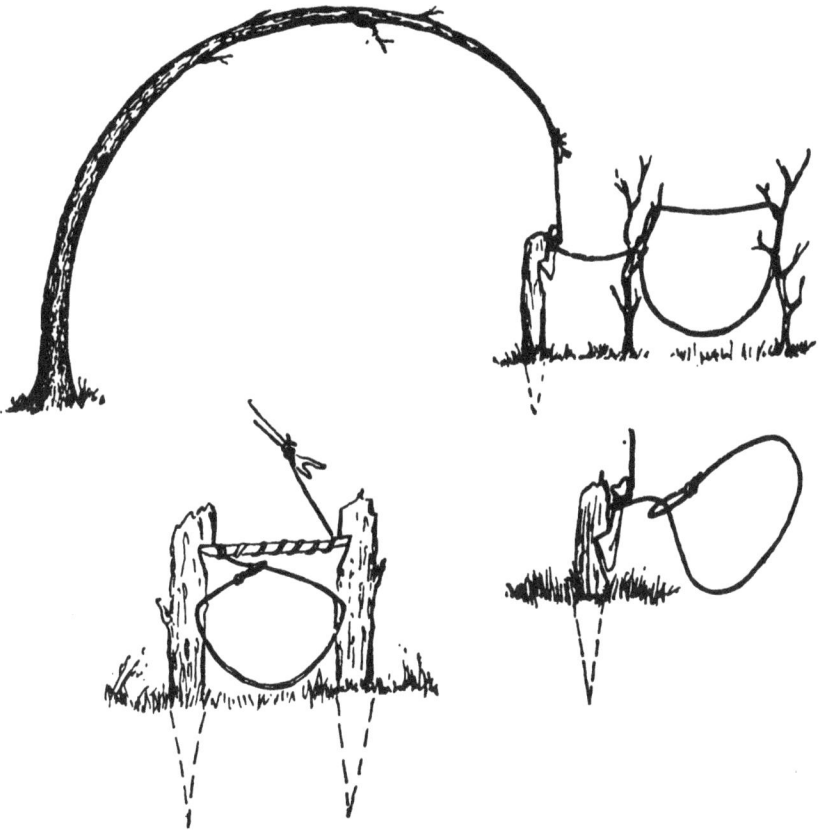

Figure 127. Fixed snare.[7]

(7) *Spring and spear trap.* You can trap jungle mammals using a bamboo spring and spear snare. As the quarry strikes the cord or wire that is secured to the trigger mechanism, the trigger is released and the spear is driven by the force of the bamboo spring (fig. 129).

(8) *Deadfall.*

  (a) You can catch medium to large animals in deadfalls; however, use this method only where bigger game exist in sufficient quantities to justify the time and

[7]Ibid.

Figure 128. A treadle spring snare.[8]

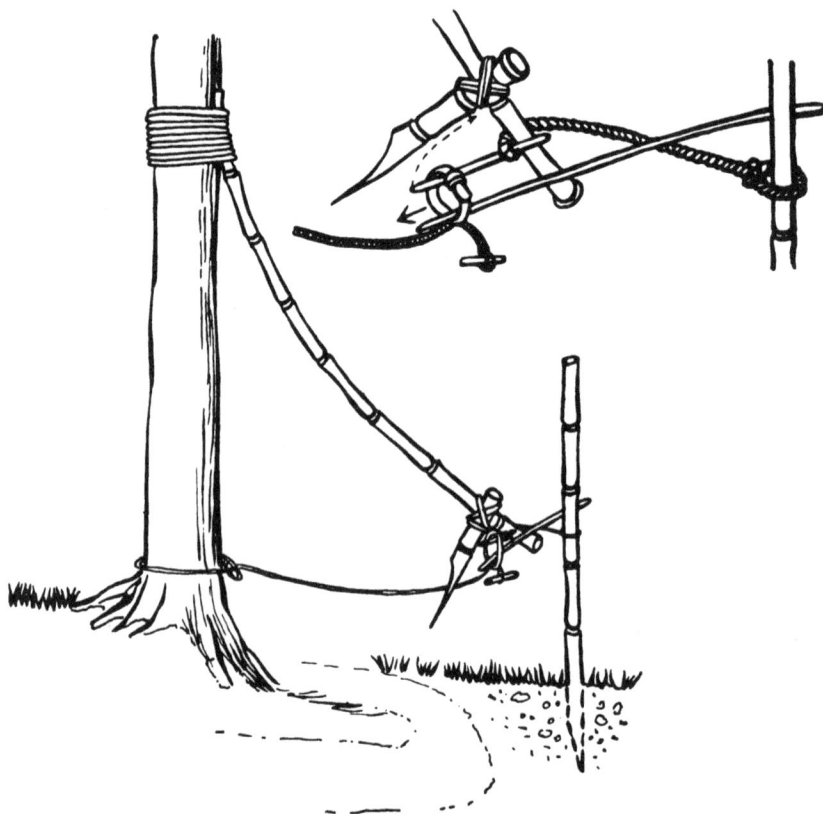

Figure 129. A spring and spear trap showing closeup of the spear and trigger.[9]

---

[8]Ibid.
[9]Ibid.

effort required to construct this snare. Build your deadfall close to or across a game trail beside a stream or on a ridge. Be sure the fall log slides smoothly between the upright guide posts, and that you place the bait far enough from the bottom log to insure time for the fall log to fall before the animal can withdraw its head (fig. 130).

(b) You can build a simple deadfall as shown in figure 131. Use a rock or heavy log and tilt it at a steep angle on a figure 4 trigger. Tie the bait on the trigger. When the game disturbs the bait, the weight will fall.

ROCK WEIGHT   PARACHUTE BUCKLE

BAIT

STOCKADE

Figure 130.   A fall log trap for big game.[10]

10 Ibid.

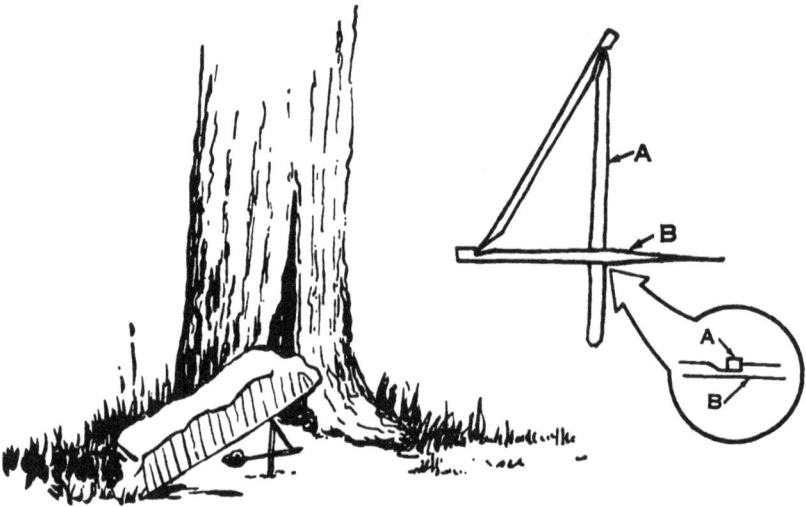

Figure 131.   A simple deadfall using a figure 4 trigger.[11]

[11] Ibid.

CHAPTER 5

FIREMAKING AND COOKING

Section I. FIREMAKING

41. Importance

a. You need fire for warmth, for keeping dry, for signaling, for cooking, and for purifying water by boiling. Survival time is increased or decreased according to your ability to build a fire when and where you need it.

b. You should be able to build a fire under any conditions of weather, if you have matches. For this reason, when operating in remote areas always carry a supply in a waterproof case *on your person.* It's advisable to learn to shield a match flame for some time in a fairly strong wind. Practice this; it could save your life.

42. Fuel, Tinder, and Location

a. Don't build your fire too big. Small fires are easier to control. Build a series of small fires in a circle around you in cold weather. They give more heat than one big fire.

b. Locate your fire carefully to avoid setting a forest fire. If the fire must be built on wet ground or snow, first build a platform of logs or stones. Protect your fire from the wind with a windbreak or reflector (fig. 132). These also concentrate the heat in the desired direction.

c. Use standing dead trees and dry dead branches for fuel. The inside of fallen tree trunks will supply you with dry wood in wet weather. In treeless areas, rely on grasses, dried animal dung, animal fats, and sometimes even coal, oil shale, or peat which may be exposed on the surface. If you are near the wreckage of an aircraft, use a mixture of gasoline and oil as fuel. Be careful how you ignite and feed the gasoline. You can use almost any plant for firewood, but do not burn the wood of any contact poisoning plant. See chapter 7.

d. Use kindling that burns readily to start your fire, such as small strips of dry wood, pine knots, bark, twigs, palm leaves,

Figure 132. Windbreak or reflector.

pine needles, dead upright grass, ground lichens, ferns, plant and
bird down, and the dry, spongy threads of the giant puffball,
which, incidently, are edible. Cut your dry wood into shavings
before attempting to set it afire. One of the best and most commonly found kindling material is punk, the completely rotted
portions of dead logs or trees. Dry punk can be found even in
wet weather by knocking away the soggy outer portions with
a knife, stick, or even your hands. Paper or gasoline may be
available as tinder. Even when wet the resinous pitch in pine
knots or dried stumps ignites readily. Loose bark of the living
birch tree also contains a resinous oil which burns rapidly.
Arrange this kindling in a wigwam or log cabin pile (figs. 133
and 134).

e. Bank your fire properly. Use green logs or the butt of a
decayed punky log to keep your fire burning slowly. Keep the
embers out of the wind. Cover them with ashes, and put a thin
layer of soil over them. Remember it takes less work to keep
your fire going than to build another one.

f. On polar ice, or in areas where other fuels are unavailable,

Figure 133. Tinder pile[12].

Figure 134. Log cabin pile.

blubber or other animal fat is a source of fuel. Animal dung may be your only fuel in desert areas. See chapter 6.

43. Fires Without Matches

a. Preparation. Prepare some extremely dry tinder before attempting to start a fire without matches. Once prepared, shelter this tinder from the wind and dampness. Some excellent tinders are punk, lint from cloth, rope or twine; dead palm frond; finely shredded dry bark; dry powdered wood; bird nests; wooly materials from plants; and wood dust produced by insects and often found under the bark of dead trees. If you want to save some tinder for future use, store it in a waterproof container.

12Ibid.

b. Sun and Glass. A camera lens, a convex lens from a binocular, or lens from a telescopic sight or flashlight may be used to concentrate the rays of the sun on your tinder (fig. 135).

Figure 135. Sun and glass.

c. Flint and Steel (fig. 136). If available, this is the best method to start tinder burning if you don't have matches. Use the flint fastened to the bottom of your waterproof match case. A hard piece of stone will serve as a substitute. Hold the flint as near the tinder as possible and strike it with a knife blade or other small piece of steel. Strike downward so that the sparks will hit in the center of the tinder. When the tinder begins to smolder, fan or blow it gently into a flame. Gradually add fuel to your tinder, or transfer the burning tinder to your fuel pile. If you don't get a spark with the first rock, throw it away and try another.

d. Wood Friction. Since the use of friction is a difficult method of starting a fire, use it only as a last resort.

    (1) *Bow and drill* (fig. 137). Make a strong bow strung loosely with a shoelace, string, or thong. Use it to spin a dry, soft shaft in a small block of hardwood. This forms a black powdered dust which eventually catches a spark. When smoke begins to rise, you should have enough spark to start a fire. Lift the block and add tinder.

Figure 136.   Flint and steel.

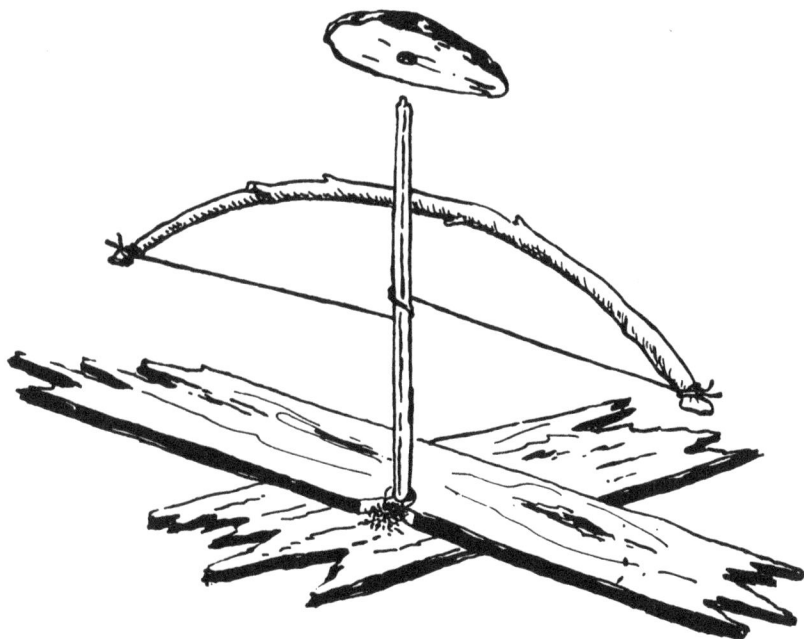

Figure 137.   Bow and drill.

(2) *Fire thong* (fig. 138). Use a strip of dry rattan, preferably about one-fourth inch in diameter and about two feet long; and a dry stick. Prop this stick off the ground by using a rock. Split the end of the stick that is off the ground and hold it open with a small wedge. Place a small wad of tinder in the split, leaving enough room to insert the thong behind it. Secure the stick with your foot, and work the thong back and forth.

(3) *Fire saw* (fig. 139). The fire saw consists of two pieces

Figure 138. Fire thong.

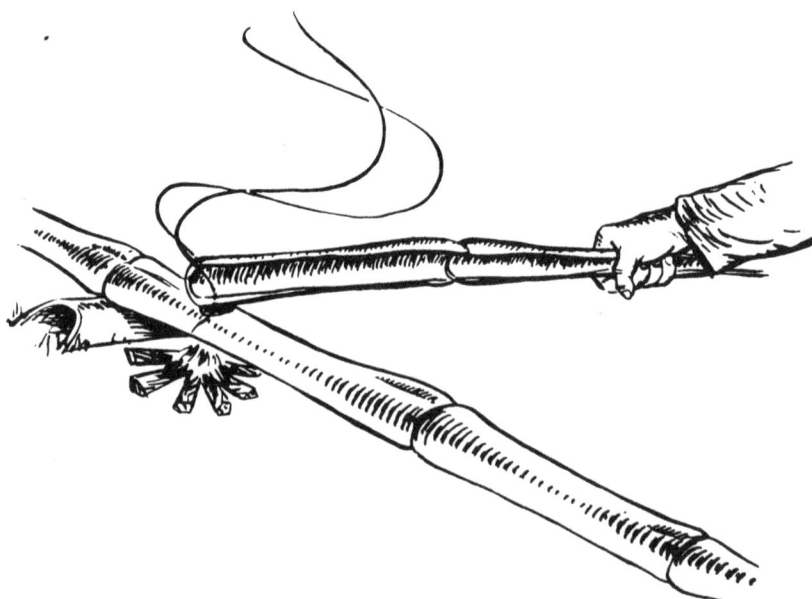

Figure 139. Fire saw.

of wood which are sawed vigorously against each other. This method of starting fires is commonly used in the jungle. Use split bamboo or other soft wood as a rub stick and the dry sheath of the coconut flower as the wood base. A good tinder is the fluffy brown covering of the apiang palm and the dry material found at the base of the coconut leaves.

e. Ammunition and Powder. See chapter 6.

44. Warming Fires

A small fire is good for warming because it requires little fuel, is easily controlled, and not so hot as to prevent your hovering over it. To realize the most heat from a small fire, drape your coat behind it or build a reflector to direct the heat. A circle of small fires keeps you more comfortable than one big fire.

45. Cooking Fires

a. A small fire and some type of stove is best for cooking purposes. Place the firewood crisscross and allow it to burn down to a uniform bed of coals. Make a simple fireplace by using two logs, stones, or a narrow trench on which to support a vessel over the fire (fig. 140).

Figure 140. Simple fireplace.

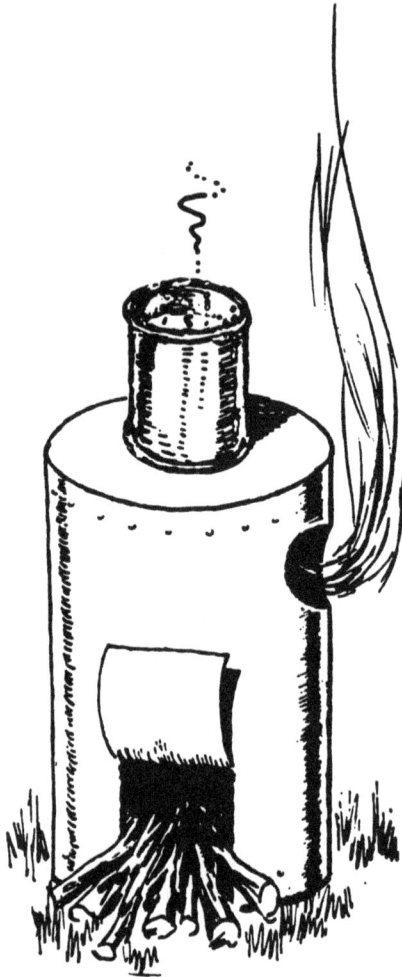

Figure 141.   Hobo stove.

b. A hobo stove made out of a tin can conserves fuel and is particularly suited to the Arctic (fig. 141).

c. A simple crane propped on a forked stick will hold a cooking container over a fire (fig. 142).

d. A bed of hot coals provides the best cooking heat. If you make a crisscross fire, the coals settle uniformly.

e. A fire that is to be used for baking should be built in a pit and allowed to burn into a bed of coals (fig. 143).

Figure 142. Simple crane.

Figure 143. Pit fire.

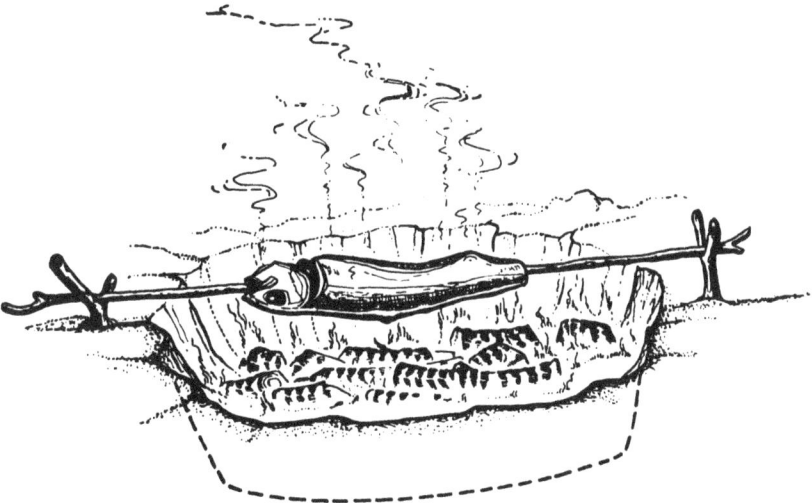

Section II. COOKING WILD FOOD

46. Skinning and Cleaning

  a. Fish. As soon as you catch a fish cut out the gills and large blood vessels that are next to the backbone. Scale it. Gut the fish by cutting open its stomach and scraping it clean (fig. 144). Cut off the head unless you want to cook the fish on a spit. Fish like catfish and sturgeon have no scales. Skin them. Small fish under four inches require no gutting, but should be scaled or skinned.

  b. Fowl.

  (1) Most fowls should be plucked and cooked with the skin on in order to retain its food value. After the bird is plucked, cut off the neck close to the body and clean out

BLEEDING

SCALING

GUTTING

SKINNING

Figure 144. Dressing fish.

the insides through the cavity. Wash it out with fresh, clean water. Save the neck, liver, and heart for stew. It is easier to pluck a fowl after scalding it (fig. 145). Waterfowl are an exception. They are easier to pluck dry.

FEATHERING

CUTTING AND GUTTING

Figure 145. Cleaning a fowl.

    (2) Scavenger birds like vultures and buzzards should be boiled for at least 20 minutes before you cook them. This kills parasites.

    (3) Save all feathers. You may want to use them for insulating your shoes or clothing or for bedding.

c. *Animals.*

    (1) *Skinning and dressing* (fig. 146). Clean and dress the carcass as soon as possible after death because to delay will make your job harder. To prepare light and medium sized animals—

        (a) Hang the carcass head downward from a convenient limb. Cut its throat and allow the blood to drain into a container. Boil it thoroughly. It is a valuable source of food and salt.

(b) Make a ring cut at the knee and elbow joints and a "Y" cut down the front of each of the hind legs and down the belly as far as the throat.

(c) From the belly make a cut down each foreleg.

(d) Make a clean circular cut around the sex organs.

(e) Working from the knee downward, remove the skin.

(f) Cut open the belly. Pin the flesh back with wooden skewers, and remove the entrails from the windpipe upward, clearing the entire mass with a firm circular cut to remove the sex organs.

(g) Save the kidneys, liver, and heart. Use the fat surrounding the intestines. All parts of the animal are edible, including the meaty parts of the skull such as the brain, eyes, tongue, and fleshy portions.

(h) Throw away the glands and entrails in the anal and reproductive regions.

(i) Save the skin. It is light when dried and is good insulation as a bed cover or article of clothing.

(2) *Larger animals.* To prepare, follow the steps outlined

WHERE TO MAKE PRELIMINARY CUTS (FOLLOW DOTTED LINES)

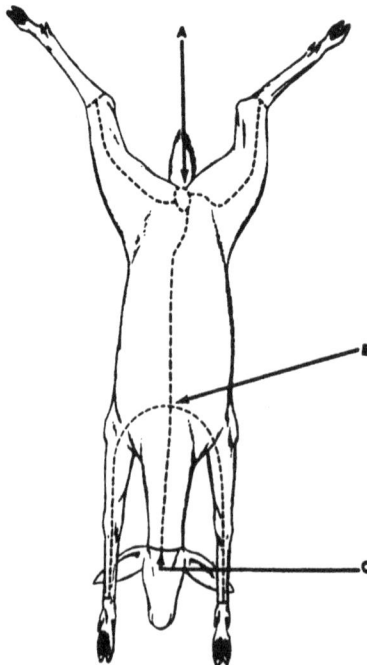

Figure 146. Dressing a carcass.

in 1(a) through (i) above, except for hanging the carcass. This may be impossible because of the lack of a suitable method of hoisting the animal.

(3) *Rats and mice.* Both rats and mice are palatable meat, particularly if cooked in a stew. These rodents should be skinned, gutted, and boiled. Rats and mice should be boiled about 10 minutes. Either may be cooked with dandelion leaves. Always include the livers.

(4) *Rabbits.* Rabbits are tasty but provide no fats to a diet. They are easy to trap and kill. To skin, make an incision behind the head or bite out a piece of skin to allow you to insert your fingers. Peel back the hide. To clean it, make an incision down the belly, spread open, and shake strongly. Most of the intestines will fall out. What remains can be scraped and washed out.

(5) *Other edible animals.* Dogs, cats, hedge hogs, porcupines, and badgers should be skinned and gutted before cooking. Prepare them as a stew with a quantity of edible leaves. Dog and cat livers are especially valuable.

d. *Reptiles.* Snakes (excluding the sea snakes) and lizards are edible. Remove the head and skin before eating. See chapters 4 and 7. Lizards are found almost everywhere, especially in tropical and subtropical regions. Broil or fry the meat.

47. How To Cook

a. *Why Cook?* Cooking makes most foods more tasty and digestible, and destroys bacteria, toxins, and harmful plant and animal products.

b. *Boiling.*

(1) *General.* When meat is tough, or when other foods require long cooking, boiling is the best way to prepare it for later frying, roasting, or baking. Boiling is probably the best method of cooking because it conserves the natural juices of the food. Remember that boiling is difficult in high altitudes and is impractical at altitudes in excess of 12,000 feet.

(2) *Vessels for boiling.* Water can be boiled in vessels made of bark or leaves, but such containers burn below the waterline unless the vessel is kept moist or the fire kept low. Half a green coconut or a section of bamboo cut well above and just below a joint can be used as containers for boiling. They will not burn until after the water boils (fig. 147). Birchbark and banana leaves make good containers (fig. 148). Secure the sides with thorns or slivers of wood. Water can be boiled in a

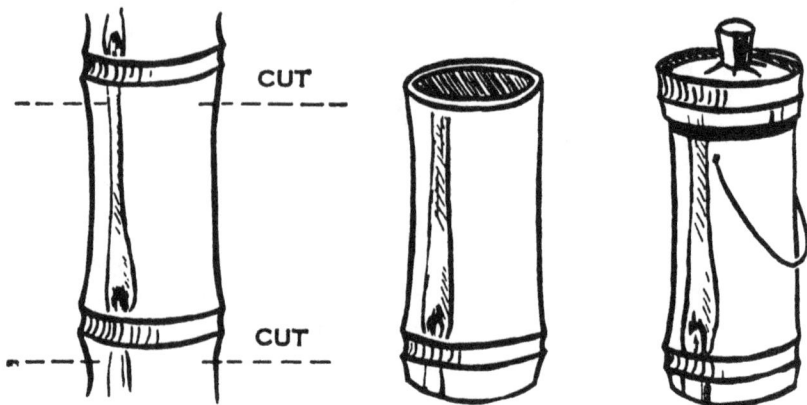

Figure 147. Bamboo container.

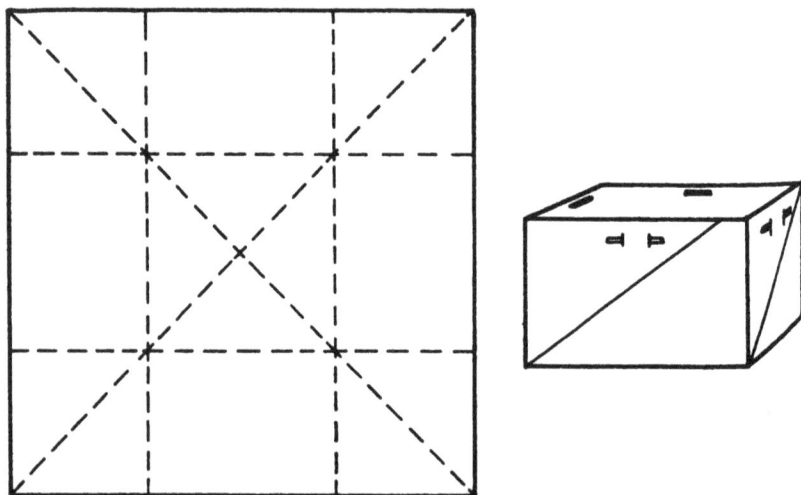

Figure 148. Birchbark vessel.

scooped out hole in clay or in a hollow log by dropping heated stones into it.

c. *Roasting or Broiling.* This is a quick way to prepare wild plant foods and tender meats. Roast meat by putting it on a stick and holding it near embers. A crane (fig. 142) may be used. Roasting hardens the outside of the meat and retains the juices.

d. *Baking.* Baking is cooking in an oven over steady, moderate heat. The oven may be a pit under your fire, a closed vessel, or leaf or clay wrapping. To bake in a pit, first fill it with hot coals. Drop the covered vessel containing water and food in the pit. Place a layer of coals over it and cover with a thin layer of dirt. If possible, line your pit with stones so that it holds more heat. Pit cooking protects food from flies and other pests and reveals no flame at night.

e. Steaming. Steaming can be done without a container and is suitable for foods that require little cooking, like the shellfish. Place your food in a pit filled with heated stones over which leaves are placed. Put more leaves over your food. Then force a stick through the leaves down to the food pocket. Pack a layer of dirt on top of the leaves and around the stick. Remove the stick and pour water to the food through the hole that remains. This is a slow way to cook, but it is effective (fig. 149).

f. Parching. Parching may be a desirable method of preparing some foods, especially grains and nuts. To parch, place the food in a metal container and heat slowly until it is thoroughly scorched. In the absence of a suitable container, a heated, flat stone may be used.

g. Utensils. Anything that holds food or water may be used as a container—turtle shells, sea shells, leaves, bamboo, a section of bark.

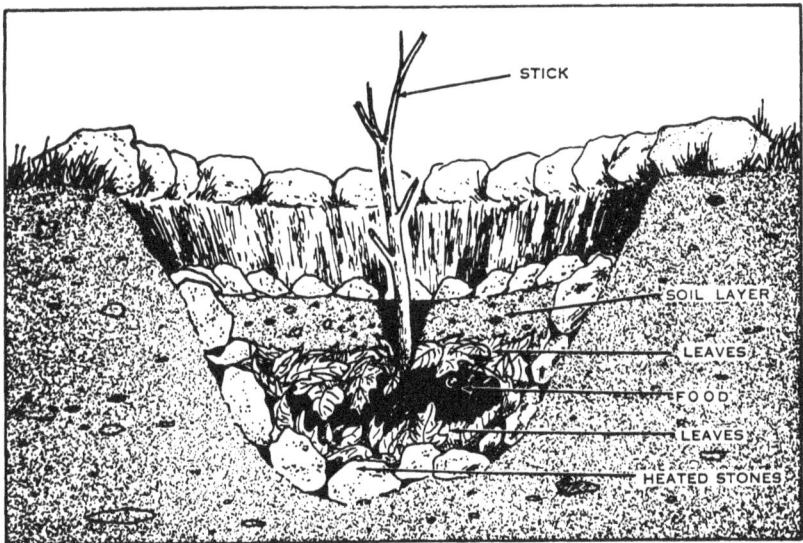

Figure 149. Steaming pit.

h. Cooking Plant Food.
    (1) *Pot herbs.* Boil leaves, stems, and buds until tender. If the food is bitter, several changes of water will help to eliminate the bitterness.
    (2) *Roots and tubers.* They can be boiled but are more easily baked or roasted.

(3) *Nuts.* Most nuts can be eaten raw, but some, such as acorns, are better crushed than parched. Chestnuts are good roasted, steamed, or baked.

(4) *Grains and seeds.* Grains and seeds are more tasty when parched but they can be eaten raw.

(5) *Sap.* You can dehydrate to a syrup any sap containing sugar. Simply boil away the water.

(6) *Fruit.* Bake or roast tough, heavy-skinned fruits. Boil succulent fruits. Many fruits are good raw.

i. *Cooking Animal Food.*

(1) *General.* Boil animals larger than a domestic cat before roasting or broiling them. Cook the meat as fast as possible when broiling because it toughens over a slow fire. When cooking larger animals, cut them into small pieces. If the meat is exceptionally tough, stew it with vegetables. If you intend to broil or bake any type meat, use some fat whenever possible. In the case of a bake, put the fat on top so that it melts and runs down on the bake.

(2) *Small game.* Small birds and mammals can be cooked whole or in part, but you should remove entrails and sex glands before cooking. Wrap a big bird in clay and bake it. The clay removes the feathers when it is broken from the cooked carcass. Boiling is the best method of cooking small game because there is less waste. Add taste to the bird by stuffing it with coconut, berries, grains, roots (onions), and greens.

(3) *Fish.* Fish may be roasted on an improvised grill of green sticks or baked in leaves and clay, or they may be cooked over direct heat by using a crane.

(4) *Reptiles and amphibians.* Frogs, small snakes, and lizards can be roasted on a stick. Large snakes and eels are better if boiled first. Boil turtles until the shell comes off. Cut up the meat and mix it with tubers and greens to form a soup. Salamanders, roasted on a stick, are edible. Skin all frogs and snakes before cooking as the skin may be toxic.

(5) *Crustaceans.* Crabs, crayfish, shrimps, prawns, and other crustaceans require cooking in order to kill disease producing organisms. They spoil rapidly, however, and should be cooked immediately after capture. Cook them alive by dropping them in boiling water.

(6) *Mollusks.* Shellfish can be steamed, boiled, or baked in the shell. Shellfish make an excellent stew with greens or tubers.

(7) *Insects.* Grasshoppers, locusts, large grubs, termites, ants, and other insects are easy to catch and will provide you nourishment in an emergency. See chapter 4.

(8) *Eggs.* Edible at all stages of embryo development, eggs are among the safest of foods. You can hard boil eggs and carry them for days as reserve food.

j. Seasoning. Salt can be obtained by boiling seawater. The ashes of burned nipa palm boughs, hickory, and some other plants contain salt that can be dissolved out in water. When the water has been evaporated, the salt has a black tint. The citric acid in limes and lemons can be used to pickle fish and other meat. Dilute two parts of fruit juice with one part salt water. Allow the fish or meat to soak for half a day or longer.

k. Baking Bread. Bread may be made with flour and water. If possible, use sea water for the salt. After kneeding the dough well, place it in a sand-lined hole. Then place sand on top of the hole and cover the dough with glowing coals. By experimentation you should be able to get the dough and temperature correct enough to prevent sand from clinging to the cooked bread. Another method of baking bread is by twisting it around a green stick from which the bark has been removed, and placing it over a fire. The stick should be bitten first to determine if the sap is so sour or bitter that it will affect the taste of the bread. Bread also may be made by spreading dough into thin sheets on a hot rock. A little leaven (dough allowed to sour) added to your bread dough improves the loaf.

48. Preserving Foods

a. Freezing. In cold climates preserve your excess foods by freezing. See chapter 6.

b. Drying. Plant food can be dried by wind, sun, air, or fire, or any combination of these four. The object is to get rid of the water.

(1) Cutting meat across the grain in one-fourth inch strips and either drying it in the wind or smoke will produce "jerky." Put the strips of meat on a wooden grate (fig. 150) and dry until the meat is brittle. Use willow, alders, cottonwood, birch, and dwarf birch for firewood because pitch woods such as pine and fir make the meat unpalatable. A parateepee (ch. 6) makes a good smoking house when the flaps at the top are closed. Hang the meat high and build a slow smouldering fire under it. Perhaps a quicker way of smoking meat is by the following method: Dig a hole in the ground about 1 yard deep and one-half yard wide (fig. 151). Make a

small fire at the bottom of the hole (after starting the fire use green wood for smoke). Place an improvised wooden grate about three-fourths of a yard up from the bottom. Use poles, boughs, leaves, or any available material to cover the pit.

(2) The methods of preserving fish and birds are much the same as for other meats. To prepare fish for smoking, cut off the heads and remove the backbones. Then spread the fish flat and skewer in that position. Thin willow branches with bark removed make good skewers (fig. 152). Fish also may be dried in the sun. Hang them from branches or spread them on hot rocks. When the meat dries, splash it with sea water to salt the outside. Don't keep sea food unless it is well dried and salted.

(3) Plantains, bananas, breadfruit, leaves, berries, and other wild fruits can be dried by air, sun, wind, or fire, with or without smoke. Cut fruit into thin slices and place in the sun or before a fire. Mushrooms dry easily and may be kept indefinitely. Soak them in water before using.

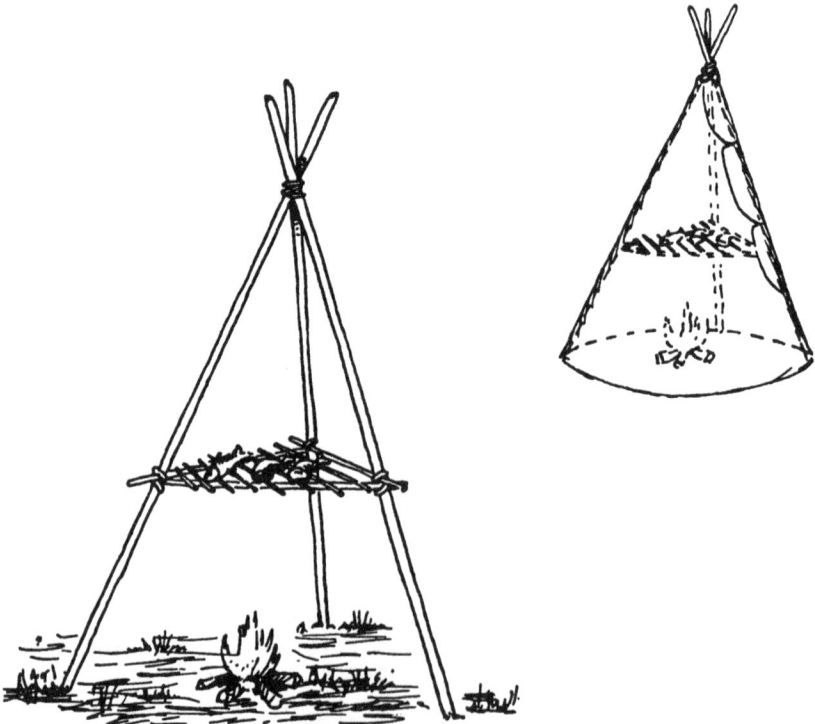

Figure 150. Wooden drying grate.

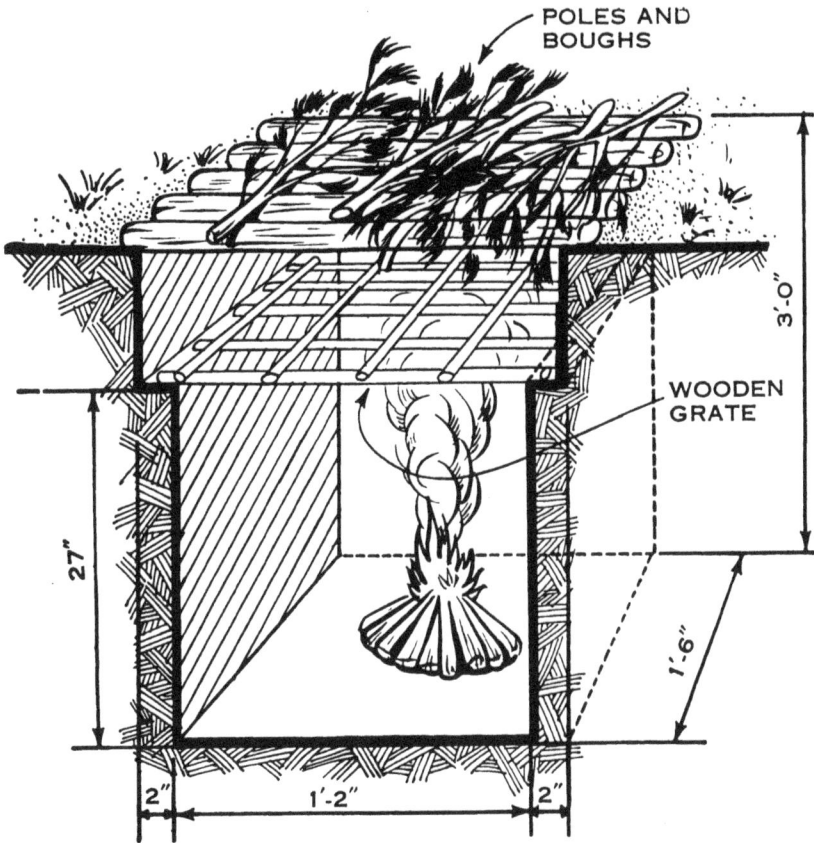

Figure 151. A pit for smoking meat.

49. Concentrated Foods

A good concentrated food is pinole. It is made by parching corn grains or seeds in hot ashes or heated stones, or in an oven. Pinole keeps indefinitely, contains a maximum of calories for its weight, is easy to prepare, and can be eaten raw or cooked. A small handful of pinole in a cup of cold water has a pleasant flavor and is highly nutritious.

50. Poisonous Plant and Animal Foods

There are relatively few poisonous plants and animals. Learn the ones that are edible and don't worry about the rest. There are some parts of animals poisonous to man, such as polar bear liver, but this and other poisonous foods are covered in chapter 7. Plants that are poisonous to eat and touch and certain poisonous sea foods are also identified in chapter 7.

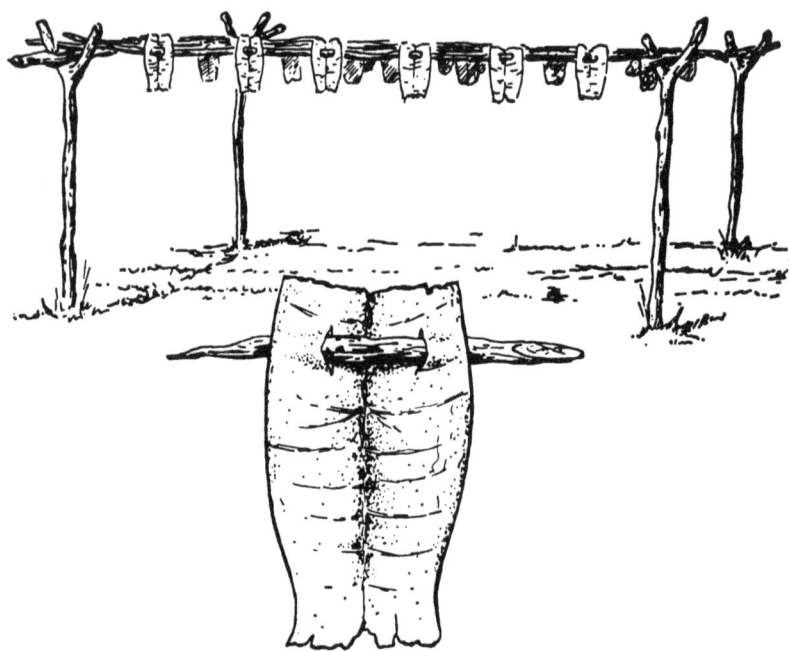

Figure 152.   Fish skewers.

CHAPTER 6

SURVIVAL IN SPECIAL AREAS

Section I. GENERAL CONSIDERATIONS

51. Know the Facts

a. Your job is to "get back." The basic principles of survival discussed in the preceding chapters will help, but your survival under extremely adverse conditions may depend upon how well these principles have been supplemented by specific information about your particular area. The more you know about conditions peculiar to your region, including the plant and animal life, the better your chances for survival.

b. Survival in remote and desolate areas, in the Arctic, desert, or jungle, depends on you. *You* must be physically fit; have a fundamental knowledge of woodcraft principles; know what foods are available and how to find and prepare them; understand how to care for your body and how to conserve energy; and recognize those plants and animals that will harm you. Armed with this knowledge, you are prepared to wage a winning battle for survival.

52. Natives Might Help

a. General. With few exceptions, natives are friendly. They know the country, its available water and food, and the way back to civilization. Be careful not to offend them. They may save your life.

b. Enlist Native Aid. To enlist native help, use these guides—
  (1) Let the natives contact you. Deal with the recognized headman or chief to get what you want.
  (2) Show friendliness, courtesy, and patience. Don't show fright; don't display a weapon.
  (3) Treat natives like human beings.
  (4) Respect their local customs and manners.
  (5) Respect personal property, especially their women.
  (6) Don't take offense at pranks played on you. Primitive people especially are fond of practical jokes.

(7) Learn all you can from natives about woodcraft and getting food and drink. Take their advice on local hazards.

(8) Avoid physical contact without seeming to do so.

(9) Paper money is worthless in many places. Hard coin is good. Also, items such as matches, tobacco, salt, razor blades, empty containers, or cloth may be valuable bartering items. One word of caution—don't overpay.

(10) Whatever you do, leave a good impression. Other men may need this help.

Section II. COLD WEATHER AREAS

53. Climate and Weather

a. *Temperature.*

(1) *Arctic.* During the Arctic summer, temperatures above 65° F. are common except on glaciers and frozen seas. Temperatures in the Arctic winter sometimes reach –70° F.

(2) *Subarctic.* Subarctic summers are short with temperatures ranging above 50° F., and occasionally reaching 100° F. Subarctic winters are the coldest in the Northern Hemisphere, ranging to extremes of from –60° F. to –80° F. in North America, and even lower in Siberia.

b. *Extremes of Temperature.*

(1) *Cold.* Extremely cold temperatures occur in the Subarctic far from the sea and at a low elevation.

(a) One of the coldest spots in the world is in the Yana River basin of Siberia, 200 miles south of the Arctic circle and 750 miles inland. Temperatures have dropped to –90° F., which is more than 120° below freezing.

(b) The coldest temperature recorded in Alaska and Canada was about —81° F., at Snag, Yukon Territory, 350 miles south of the Arctic Circle and hundreds of miles inland.

(2) *Heat.* Very warm temperatures are also found in the Subarctic, far from the moderating influence of the sea. The highest recorded temperature for Alaska occurred at Fort Yukon where a temperature of 100° F. in the shade was reported. High temperatures are also found inland along the Mackenzie River in Canada and along the north-flowing rivers of Siberia.

c. *Winds.*

(1) *General.* Winds during the summer blow inland from

the Arctic Ocean. In the winter prevailing winds blow outward from the cold centers in Siberia, Greenland, and northwest Canada. Winds are strongest in the Arctic tundra and mountain areas, less strong in the forested Subarctic. Great drifts of snow form in the shelter of any object. The most noticeable effect of the wind in summer is to blow away the multitude of annoying insects.

(2) *Wind-chill factor.* In winter, the wind when accompanied by low temperatures chills man quickly. The measure of loss of heat is called the "wind-chill factor." Wind-chill is the combined cooling effect of air, temperature, and wind on a heated body, rather than the temperature as recorded by a thermometer.

(3) *Winds and snow.* Winds are disagreeable enough when associated with cold, but when they are also accompanied by drifting snow, they are tortuous and at times make traveling almost impossible. However, the Subarctic forest naturally decreases the effect of the wind and affords some cover. Winds of 9 to 12 miles per hour will raise the snow a few feet off the ground. Winds of 15 miles per hour, or more, will raise snow high enough to obscure buildings. At 30 miles per hour, and above, snow may be whisked up to 50 to 100 feet and resemble a low cloud.

(4) *Gales.* Violent local gales occur frequently in the Subarctic and Arctic. Near land, especially where there is a plateau descending to the sea, there are strong gales. This is true on the slopes of Greenland, particularly on the west coast. In autumn and early winter fierce gales are common on the north coast of Canada, around Franklin Bay and Herschel Island. The famous Aleutian "williwaw", springing up suddenly and changing direction quickly, is a striking example of Arctic winds and gales that frequently reach hurricane velocity.

d. *Precipitation.*

(1) Many areas of the Far North receive less precipitation in the form of rain or snow than the dry southwestern United States. Snow remains on the ground throughout the winter because of low temperatures. In summer little moisture is lost through evaporation. The average annual precipitation in the Subarctic except near seacoasts is the equivalent of 10 inches of rainfall, while in the Arctic it is generally 5 inches or less.

(2) Fogs are found over open water and over the ice area

near the ring of open water. During the winter months —December through March—you find practically no fog at all when you are more than 200 miles from the nearest open water. Fogs increase in frequency from April to July, in which months about every other day is foggy. Toward the end of August, and from September on, the fogs clear up.

(3) The Arctic coastal region is a land of summer fog and dense cloud cover, with clearing conditions accompanied by high winds of biting cold during the winter months. The dense coastal fog begins in May and recurs persistently until August. Summer fog is chiefly marine fog which is carried over the land by on-shore winds. Overhanging fog banks are found in valleys and basins which open on the water. Another type of fog occurs when cold wind blows in from the frozen sea over warm, moist land. The cold air is quickly saturated from below and fog forms a short distance on shore. This fog is most dense several miles from the coast and may extend 30 miles inland. In these two cases, strong winds dissipate fogs and produce low cloudiness instead.

e. *Landscape.*

(1) Landscape varies very widely in Subarctic and Arctic lands including practically every gradation between mountain peaks and glaciers to the flattest of plains. Greenland, Ellesmere, Devon and Baffin Islands, Iceland, Spitzbergen, Novaya Zemlya, and Severnaya Zemlya are all rugged islands with ice caps and mountainous interiors. On the North American mainland, extremely wild mountain country occurs in the west in the Alaskan, Brooks, and Richardson Mountains, and in the east in Labrador. Similar areas occur in northeast and north Siberia, and in the Scandinavian Peninsula. Vast plains are found in Canada, Alaska, and Siberia, notably in the basins of the major rivers. But within these plains are low, hummocky, and rocky areas of hills and ridges separated by swamps, bogs, rivers, and lakes.

(2) Summer surface conditions in both the Arctic and Subarctic also include practically every gradation between the extremes of the hardest and roughest surfaces to the softest and wettest surfaces. Bare rock, boulder fields, gravel, sand or clay, and swamps, bogs, and lakes all occur both as extensive areas and mixed together in every conceivable combination.

(3) In winter the lakes, rivers, and swamps are frozen and become the highways of the North. Snowfall smooths out and improves movement over areas of minor surface roughness, but is too light to overlay completely and to neutralize boulder fields and rough rock surfaces.

54. What Are Your Chances?

Your odds of surviving in these areas of extremes are better than you think. The proper attitude—a will to survive— and a few elementary precautions will increase your chances. Learn to work with nature, not against it.

55. Travel

a. Once evasion becomes necessary, you must lay plans of movement carefully and adhere to them. The secret of successful travel in cold weather areas is sufficient food and rest, *and a steady pace.*

b. Your course should be determined by your location and the terrain. In mountainous or wooded areas it is advisable to follow rivers downstream toward populated areas. Siberia, where rivers flow northward, is an exception. Populous areas lie south in Siberia and European Russia; there are only a few scattered natives northward in these areas.

c. When traveling across country, try to follow the contour of the land. However, note that valley floors are frequently colder than slopes and ridges, especially at night. Head for a coast, major river, or known point of habitation.

d. During the Arctic winter, there are five basic requirements that you must meet to travel successfully—

(1) Knowledge of your exact location of departure and the location of your objective (ch. 2).

(2) Knowledge of methods of confirming your course (ch. 2). In the true Arctic, the North Star is too high in the sky to help in determining direction. Other constellations must be used, or a very accurate determination can be made by shadows. Hang a rock from the end of a stick propped at a 45° angle to the ground (fig. 153). At some time before noon mark the spot where the rock's shadow falls. Approximately six hours later mark the spot where the afternoon sun casts the rock's shadow. Draw a line from the point directly under the suspended rock through a point halfway between the morning and afternoon marks. This line points to within 3° of true north. Arctic "visual aids" can be used to determine direction. For instance snow

drifts usually are on the lee or downwind side of protruding objects like rocks, trees, clumps of willows, or high banks. By determining the cardinal points of the compass and from them the direction of the drifts, the angle at which you cross them will serve as a check point in maintaining a course. The snow on the south side of the ridges tends to be more granular than on the north. Other aids to determining direction are willows, alders, and poplars, which tend to lean toward the south, and coniferous trees which are more bushy on the south side. Use these aids as a means of *very rough* estimation.

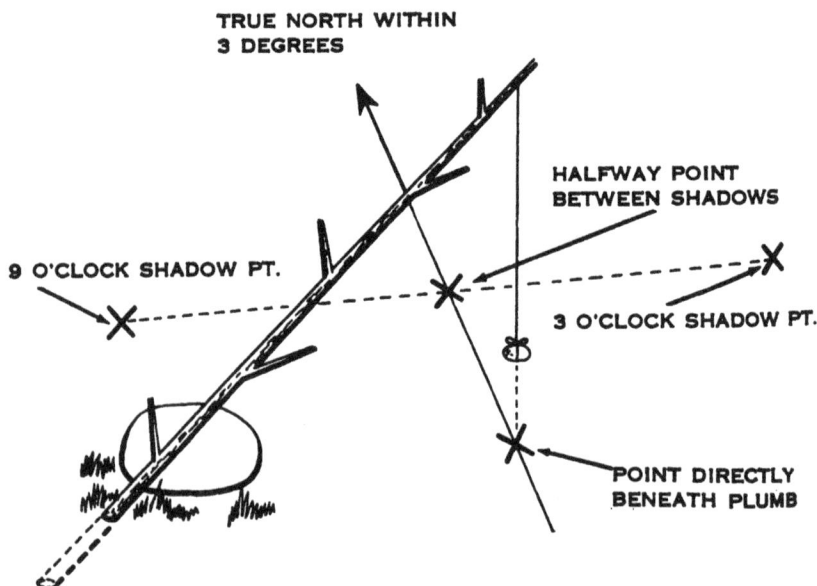

TRUE NORTH WITHIN 3 DEGREES

HALFWAY POINT BETWEEN SHADOWS

9 O'CLOCK SHADOW PT.

3 O'CLOCK SHADOW PT.

POINT DIRECTLY BENEATH PLUMB

Figure 153. Determining direction by shadows.

(3) Physical stamina. Survival is synonymous with "take your time." Without proper equipment and in poor weather, there are few people who possess sufficient stamina to travel successfully in the Arctic.

(4) Suitable clothing for the season and terrain. Remember that you must have sufficient clothing to allow you a reasonable chance of remaining dry.

(5) Food, fuel, and shelter—or equipment which permits you to obtain them from your surroundings. You require more food when traveling than when inactive. Therefore, if food is limited and little game is present

in the area to be traveled through, make certain that travel is the only solution.

e. Obstacles to summer travel are dense vegetation, rough terrain, insects, soft ground, swamps and lakes, and unfordable rivers. Winter obstacles are soft snow, dangerous river ice, severe weather, scarcity of native food, and "overflows" (stretches of water covered only by thin ice or snow). When traveling in the Arctic you should—

(1) Avoid traveling during a blizzard.

(2) Take care when crossing thin ice. Distribute your weight by lying flat and crawling.

(3) Cross glacier-fed streams when the water level is lowest. This may be at any time during the day, depending on distance from the glacier, temperature, and terrain.

(4) Take into consideration the clear Arctic air which makes distance estimation difficult. Underestimates of distances are more frequent than overestimates.

(5) Avoid travel in "white out" conditions when lack of contrast makes it impossible to judge the nature of the terrain.

(6) Always cross a snow bridge at right angles to the obstacle it crosses. Find the strongest part of the bridge by poking ahead with a pole or ice-axe. Distribute your weight by wearing snowshoes or skis or by crawling.

(7) Make camp early to have plenty of time to build shelter.

(8) Consider rivers as avenues of travel, frozen or unfrozen. When rivers are frozen, they frequently are clear of loose snow, and the smooth ice along its edges makes for easier travel.

f. Your ability to travel successfully over snow-covered terrain is directly related to the following factors:

(1) Your ability to use and the availability of oversnow equipment. If you possess some previous training in cross-country skiing and equipment is available, travel on skis is recommended. In most snow conditions and over most types of terrain, skies provide the speediest and most energy-saving mode of travel. Use of snowshoes requires hardly any previous training, but your speed will be much slower and travel more exhausting.

(2) Prevailing snow conditions. In deep, loose snow, skiing is exhausting and, if you have a choice of equipment, snowshoes are recommended. Even a light crust on the surface of the snow prevents skis from sinking and

provides for speedy and easy skiing. A crust hard enough to support a man, makes travel on foot feasible, but even then, if equipment is available and you possess the necessary proficiency, travel on skis is recommended.

(3) Improvise equipment for travel if snow is loose and deep. Make snowshoes of willow or other green wood, using a wood separator and thong, wire, cord, or shroud (fig. 154). If wreckage of an aircraft is available, make snowshoes out of seat bottoms, inspection plates, and other parts of salvage.

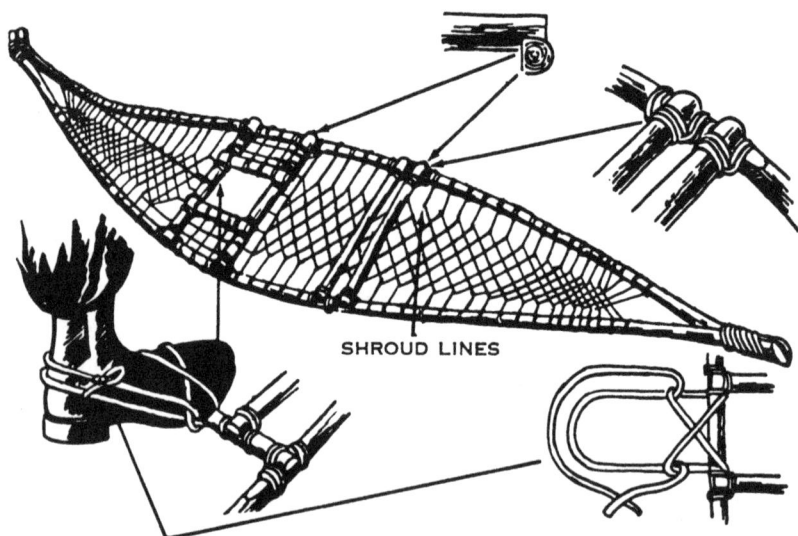

SHROUD LINES

Figure 154. Improvised snowshoes.

56. Shelter

a. Cold Can Kill. You cannot expect to survive winter cold unless you have some sort of protection. During the summer, however, you may need shelter only as a protection against insects and the sun. Suitable natural shelter may be available in caves, rock overhangs, crevices, bushy clumps, or natural terraces.

b. Selecting a Site. Ideal sites for shelter differ in winter and summer. The choice during winter depends upon protection from the wind and cold and nearness to fuel and water. In mountainous areas, you must consider the danger of avalanches, rock falls, and floods. You should choose a site during the summer months which is relatively free of insects and near fuel and water. As a protection against insects, it is better to select a site on a breezy ridge or in a place that receives an onshore breeze. Sites in forests and near rapid streams are desirable. Remember you may be in a situation where concealment is the most important factor in selecting a site for shelter; therefore, your site should afford good observation and have one or two concealed routes of escape from it.

c. *Types of Shelter.* The type shelter you build depends upon materials available and the time you have to build it. Regardless of the type, however, the Arctic shelter must serve the principal purpose of holding the heat of a fire or that of your body around you so that you remain warm. You retain your body heat longer in still air. For this reason build your shelter small, snug, and windproof. It must also provide adequate ventilation to prevent your asphyxiation. Make a hole at the top of your shelter to allow carbon monoxide gases and smoke to escape. Leave a small crack near the bottom to let in fresh air.

(1) On pack ice or snow-covered barren land, you can dig in or build up the ice and snow. Building up is sometimes easier than digging in. However, since a projecting shelter is more readily seen, it may be inadvisable due to the situation.

(2) Of several kinds of improvised shelter, perhaps the simplest is provided by a hard-packed snow drift hollowed out to accommodate one or more men. Even a hole in the snow provides temporary emergency shelter. This type of shelter sometimes is harder to make than it sounds because of the hardness of the packed snow. Frequently it is impossible to make unless you have proper tools (fig. 155).

(3) A house built of snow blocks is a useful semipermanent refuge for two or more survivors. To build a snowhouse requires considerable experience and practice, however. The emplacement of the blocks in this type structure is the important point to remember, since the blocks are supported by three impinging corners—the two bottom corners and the top (fig. 156). The support of the three corners, aided by the downward slope of the inclined plane, is the only "mystery" in snowhouse construction. Crevices between blocks are stuffed with triangular pieces of snow and finished off with soft snow gently rubbed in with a mittened hand. The snow functions as a binder and becomes stronger than the original snow blocks. A drawback in constructing this type shelter is your need for tools—knife, saw, or axe.

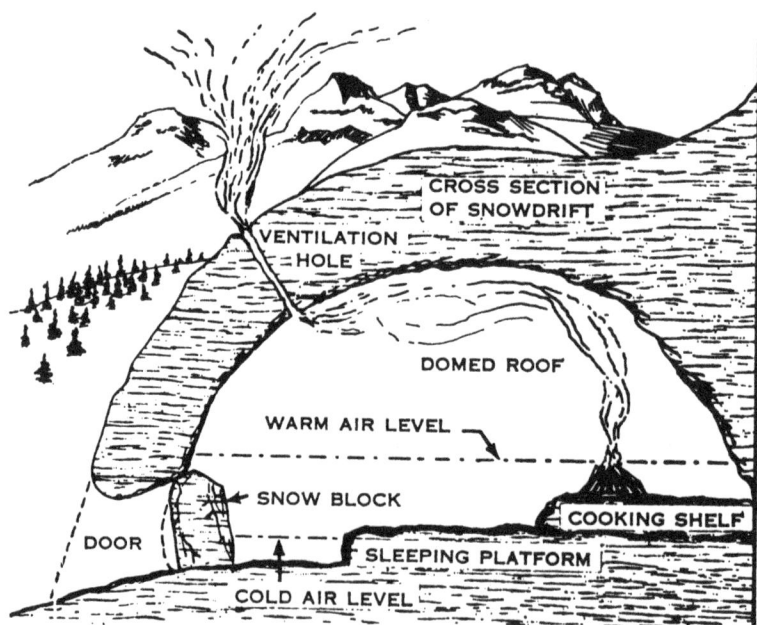

Figure 155. Snow caves.[13]

It has been proved by Eskímos that with a knife you survive—without it you need a miracle.

4) The lean-to is the standard timber shelter (fig. 157). When you use a lean-to, however, it is important that you be tactically located where you can build a fire adequately large to spread warmth equally throughout the shelter. The proper placing of the lean-to and fire in relation to prevailing winds is another consideration.

---

[13]Aviation Training, Office of the Chief of Naval Operations, United States Navy, "How To Survive on Land and Sea," Copyright 1943, 1951 by The United States Naval Institute.

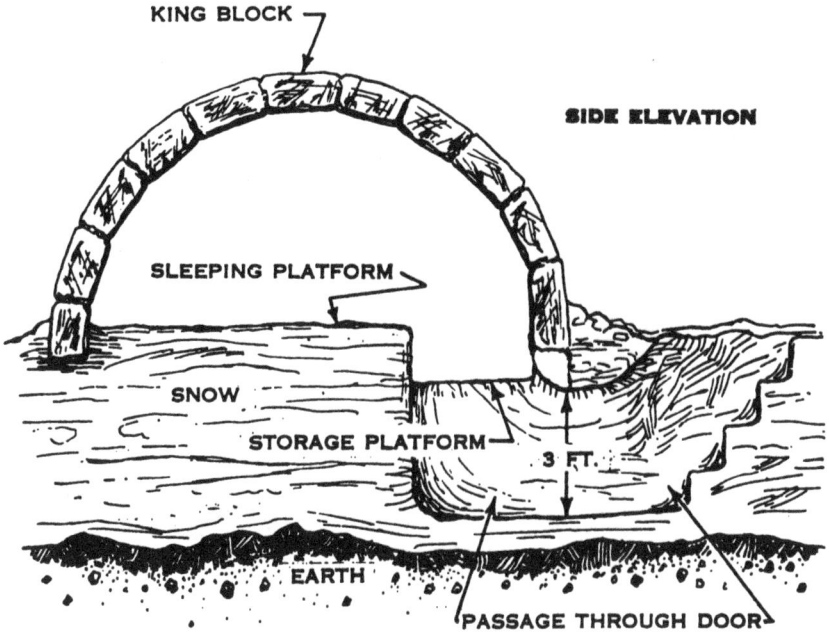

KING BLOCK

SIDE ELEVATION

SLEEPING PLATFORM

SNOW

STORAGE PLATFORM

3 FT.

EARTH

PASSAGE THROUGH DOOR

Figure 156. Snow house.

This shelter can be improved by the use of a fire and reflector (fig. 158), which is built of green logs and placed on the opposite side of the fire from the opening of the lean-to. Large stones stacked behind the fire also reflect heat.

Figure 157. Method of pitching lean-to.

Figure 158. Fire base and reflector.

(5) The parateepee is a tent made from a parachute. It is easily built and especially suited for protection against damp weather and insects. You can cook, eat, sleep, rest, and make signals without having to go outside. A number of good poles about 12 to 14 feet long are required for parateepee construction (fig. 159).

(6) A satisfactory shelter can be made by tying willow trees together to form a framework which can be covered by fabric. There is no particular design for this type of shelter, but it should be large enough for one man and his equipment. Place the open end of the willow shelter at right angles to prevailing winds. Pack the edges of the cover down with snow to prevent wind from blowing under it (fig. 160).

(7) Shelters made of boughs do not reflect the heat of a fire and become very undesirable during a rain. But boughs may serve as suitable temporary shelter (fig. 161).

(8) A log shelter is quickly and easily built. Place two poles on a large log, and cover the frame with foliage (fig. 162). This shelter is not a suitable permanent one.

d. *Beds.*

(1) When you complete your shelter, build a comfortable

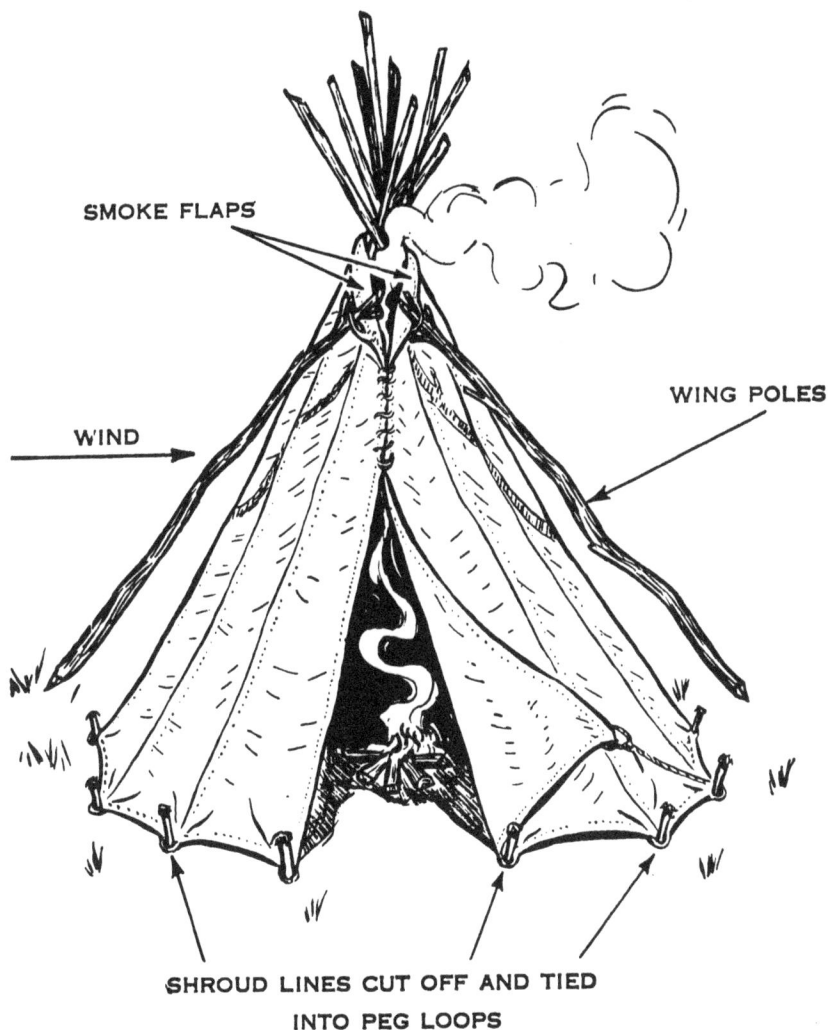

SMOKE FLAPS

WING POLES

WIND

SHROUD LINES CUT OFF AND TIED
INTO PEG LOOPS

Figure 159. Parateepee.

bed. Make it so that you will be insulated against the
cold, damp ground. If a parachute is available, spread
over a bed of leaves. First warm and dry the ground
by building a fire over the bed area, and then stamp the
hot coals into the ground. A parachute may also be
used as a hammock.

(2) Build a bed of boughs (fig. 163). Insert the branches
in the ground with the tips slanted in one direction.
Place them about eight inches apart. Cover the boughs
with fine tips.

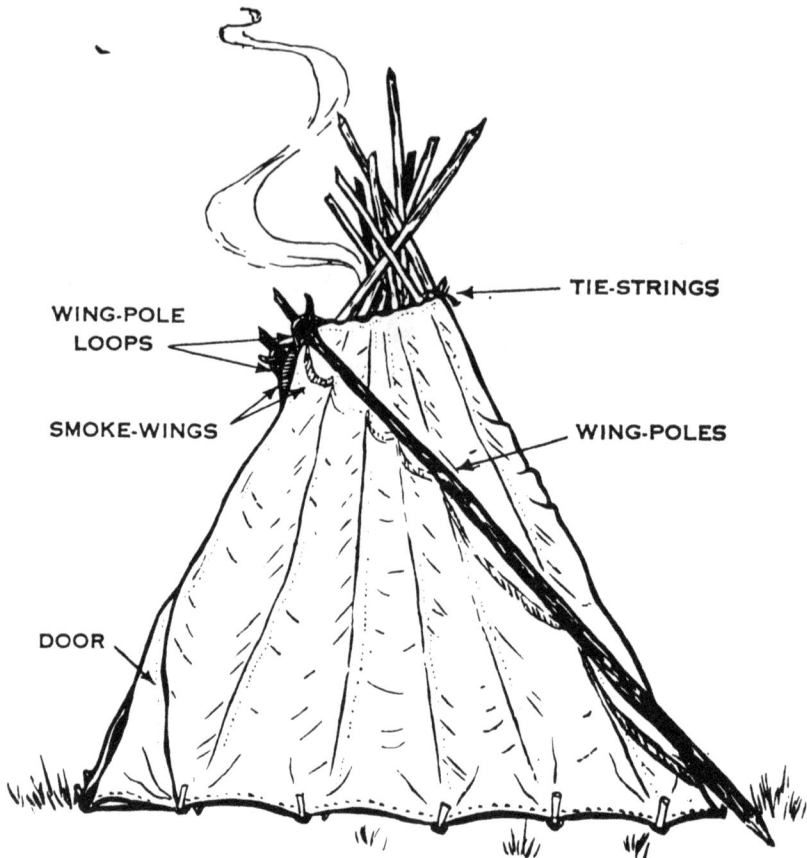

Figure 159—Continued.

57. Water

a. Thirst is a problem in cold regions during the winter. In order to conserve fuel for other purposes the survivor often deprives himself of drinking water which might have been obtained by melting ice or snow. The time and energy required to chop and gather ice for water also tends to limit the supply. A survivor may become dangerously dehydrated in cold, Arctic regions just as easily as in hot, desert areas.

b. You can obtain water by cutting a hole in the ice or by melting ice. Remember it requires approximately 50 percent more fuel and time to obtain an equal amount of water from snow than from ice.

c. It is safe within limits to eat snow, but observe these precautions—

    (1) Allow snow to thaw long enough so that you can mould it into a long "stick" or "ball." Do not eat snow in its

Figure 160.   Willow shelter.

Figure 161.   Bough shelter.[14]

14Ibid.

Figure 162.   Log shelter.[15]

Figure 163.   Bough bed.[16]

natural state—it will cause dehydration instead of relieving thirst.

(2) Do not eat crushed ice as it may cause injury to your lips and tongue.

(3) If you are hot, cold, or tired, to eat snow may lower your body temperature.

d. There are many ponds, lakes, and streams from which you may obtain water during the summer. Depressions on icebergs and floes contain fresh water during the warmer months, as do some protected coves and inlets where water from melting snow has accumulated. But all water, regardless of source, should be boiled or treated by chemical means, if such can be done. Untreated river water is dangerous. Pond water, although brownish in tint, is usually fit to drink. The milky water of a glacial stream can be drunk after sediment is strained out or allowed to settle. Old melted sea ice, recognized by its bluish color and rounded corners, is drinkable. New sea ice is too salty.

e. Any surface that absorbs the sun's heat can be used to melt ice and snow—a flat rock, dark tarpaulin, or signal panel. Ar-

[15]Ibid.
[16]Ibid.

range the surface so that the water drains into a hollow or container.

58. Food

a. *Abundance.* Your chances for finding different foods vary in the Arctic, depending on the time of year and the place. Arctic shores normally are scraped clean of all animals and plants by winter ice; but north of the timber-line, and whenever food such as mice, fish, and grubs is not available, you can still find enough food to survive.

b. *Storage and Preservation.*

(1) In the event you kill a large animal or find an abundance of smaller game, you probably will store or preserve some of your meat for future use. During cold weather, freezing fresh meat or fish preserves it. Freeze it as quickly as possible by spreading it around outside your shelter.

(2) During summer months, meat and game should be kept in a cool shady place. A hole in the ground will substitute as a refrigerator. Cure meat by hanging it in strips in trees where the wind and sun can reach it (ch. 5).

(3) In some areas it may be necessary to protect your supplies from small animals. This can be accomplished by hanging the supplies about six feet above the ground or by using wilderness caches (fig. 164).

c. *Fish.* There are few poisonous fish in Arctic waters. But some fish, like the sculpin, lay poisonous eggs; the black mussel may be poisonous at any season, and its poison is as dangerous as strychnine. Also avoid Arctic shark meat. In coastal streams and rivers, salmon moving upstream to spawn may be plentiful. But their flesh deteriorates as they travel away from the sea, making them unfit to eat except as a last resort.

(1) In the North Pacific and North Atlantic, extending northward into the Arctic Sea, the coastal waters are rich in all sea foods. Grayling, trout, white fish, and ling are common to the lakes, ponds, and the Arctic coastal plains of North America and Asia. Many larger rivers contain salmon and sturgeon. River snails or fresh water periwinkles are plentiful in the rivers, streams, and lakes of northern coniferous forests. These snails may be pencil-point or globular in shape (fig. 165).

(2) Fish can be speared, shot, netted, hooked, caught by hand, or stunned by rock or club.

Figure 164. Wilderness caches.

Figure 165. Periwinkle.

(a) Improvised hooks. See chapter 4.
(b) A piece of meat, insects, or minnows can be used for bait. Some northern fish nibble at any small object that hits the water. Cod swim up to investigate strips of cloth or bits of metal or bone. Cod may be caught through a hole in the ice.

(3) A good net can be made out of stout twine or from the inner strands of parachute shroud lines. For the full-grown salmon trout, the meshes should be about two

inches square. A scoop net with very fine mesh is required for smaller fish. This can be made out of a pliable willow branch and netting or twine. See chapter 4.

(4) Fish can be netted or clubbed more easily in a narrow part of the stream. Narrow a shallow stream by building a fence of stones, stakes, or brush from either bank.

(5) Fish are sometimes left stranded when a stream is diverted. This method may be very successful (fig. 166).

(6) To strand fish when the tide goes out, pile up a crescent of boulders on the tidal flats, scooping out the area inside.

Figure 166. Diverting a stream.

d. *Land Animals.*

(1) *Large land game.* You can expect to find deer, caribou, wild reindeer, musk ox, and moose, elk, mountain sheep and goat, and bear in Arctic and Subarctic regions. Animals of the Arctic and Subarctic are circumpolar in distribution; therefore, they are generally found in

Alaska, Northern Canada, Labrador, Greenland, Northern Europe, Iceland, and Siberia.

(2) *Small land game.* Tundra animals include rabbits, mice, lemming, ground squirrel, and fox, among others. They are found winter and summer on the tundra. Ground squirrels and marmots hibernate in the winter; during the summer squirrels are abundant along sandy banks of large streams. Marmots are found in the mountains among rocks, usually near the edge of a meadow—much like our woodchucks. Farther south where trees occur, the porcupine is often encountered and can be easily shaken from a tree and clubbed. This animal feeds on bark, and limbs stripped bare are good signs of its presence. Pick up a porcupine only after it is dead.

(3) *Hunting hints.* Hunting is generally better during the early morning and late evening when the animals are moving to and from feeding and bedding grounds and water. *Use your ammunition on large game.* Large animals in the Arctic are fairly easy to stalk and kill, and they supply you with much food and fuel. Their skins are also very useful. To successfully hunt land game, you should know some of their characteristics.

(a) Caribou or reindeer may be very curious. It is possible to attract them near enough for a shot by waving a cloth and moving slowly toward them on all fours.

(b) Wolf may be brought close by a four-legged pose.

(c) Moose may be found in heavy brush; they may charge you. In the winter you can spot them by climbing a hill or tree and looking for the animal's "smoke," (condensed body vapor which rises like the smoke of a small campfire).

(d) Mountain goat and sheep are wary and hard to approach. You may surprise them however, by moving quietly downwind while they are feeding with their heads lowered. Stay on higher ground than they are.

(e) Musk ox leave cattlelike tracks and droppings. When alarmed, they group together and remain in that position unless approached; then one or more bulls may charge you. Shoot them in the neck or shoulder.

(f) Bears can be surly and dangerous. A wounded bear is extremely dangerous and should not be followed into cover. The polar bear is a tireless, clever hunter with good sight and an extraordinary sense of smell. *He may even hunt you!*

(g) Rabbits often run in circles and return to the same

place when they are frightened. If the animal is running, whistle. It may stop. Snares are efficient for smaller land game (ch. 4).

e. *Sea Animals.* During the winter and spring sea mammals—seals, walrus, and polar bears—are found on the frozen pack ice and on floes in open water. Like land animals, these sea animals supply food, implements, fuel, and clothing.

    (1) Seals are hard to approach but can be stalked (fig. 167). Keep downwind and avoid sudden moves. A white camouflage suit helps. Advance only when the animal's head indicates it is sleeping. If a bearded seal appears ready to move off, stand up quickly and yell; the seal may become frightened and lie still, allowing you to shoot or spear it. The bearded seal stays on floe ice. The seal is found in number where ice is broken by current holes and tidewater cracks. Don't eat the liver of the bearded seal; its high vitamin A content may make you sick.

Figure 167. Seal stalking.

    (2) The walrus comes up to breathe but it is harder to locate than the seal because it does not scratch breathing holes in the ice. Walrus are found on floe ice, and generally you must approach them by boat. They are probably one of the most dangerous animals of the Arctic.

    (3) Polar bears are found in practically all Arctic coastal regions but they rarely come on land. Avoid them if possible, but if it is necessary to kill one for food, do not eat the liver. It is dangerous because of its high concentration of vitamin A. *Never eat polar bear meat unless it is cooked.* It is always diseased.

f. *Birds.*

    (1) *General.* The breeding grounds of many birds are in the Arctic. Ducks, geese, loons, and swans build their

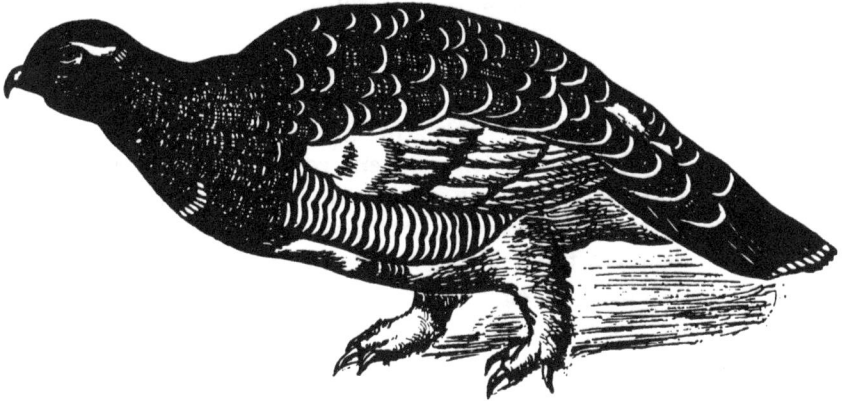

Figure 168. Ptarmigan.

nests near ponds on the coastal plains during the summer, and they provide an abundant source of food. Grouse and ptarmigan (fig. 168) are common in the swampy forest regions. Sea birds may be found on cliffs or small islands on the coast. Their nesting areas often can be located by their flights to and from feeding grounds. Sea birds, as well as ravens and owls, are useful for food.

(a) In winter, owls, ravens, and ptarmigans are the only birds available. Rock ptarmigans are easily approached, travel in pairs, and are very tame. Although hard to locate because of their protective coloring, they provide a good source of food because they can be killed with stones, slingshot—or even a club. Willow ptarmigans gather in large flocks in willow clumps in bottom lands and are easily snared (ch. 4).

(b) All Arctic birds go through a 2- to 3-week flightless period in the summer while they are moulting. When birds are moulting, you can run them down. Remember that fresh eggs are among the safest of foods, *and they are edible at any stage of embryo development.*

(2) *Bird traps.* Birds may be caught in a variety of ways—netted with an improvised net made from cord; with a baited hook attached to a fishing line; by simple traps; or, in the case of half-grown birds, by hand (ch. 4).

g. *Plant Foods.*

(1) Most plants in polar regions are edible. The water hemlock (fig. 169) is in most cases the only seriously

poisonous plant, but buttercups and some mushrooms should be avoided (ch. 4).

(2) Some of the more common edible plants include—

    (a) *Lichens* (figs. 170 and 171). Almost universal, lichens have possibly the greatest food value of all Arctic plants. Some lichens contain a bitter acid that may cause nausea and severe internal irritation if eaten raw. Soaking and boiling the plants in water removes the acid. Lichens can be prepared as a powder by soaking them overnight and allowing them to dry. They can be made crisp by roasting them slowly in a pan. Pound the dried lichen with stone and soak the lichen powder for a few hours. Then boil it until it forms a jelly. Use this to thicken soups and stew vegetables. Rock tripe (fig. 171) consists of thin, leathery, irregular shaped discs, one to several inches across. It is black, brown, or greyish. The discs are attached to rocks by a short central stalk. This lichen is soft when wet, hard and brittle when dry.

    (b) *Berries* (ch. 4). The salmonberry is the most important of the northern berries, but all berries except the baneberry are edible (fig. 172).

        1. Baneberry (fig. 173). This berry is poisonous.

        2. Mountain berry (fig. 174) is a low creeping shrub with leathery evergreen leaves. It has red berries which are high in vitamin content.

        3. Alpine bearberry (fig. 175) grows on a trailing shrub with shreddy bark and rounded leaves that turn red and are almost tasteless. Dry and shred the the leaves to make a fairly good tobacco substitute.

    (c) *Roots.* The following roots are edible:

        1. Sweet vetch (fig. 176) supplies the licorice root which is edible raw or cooked. The plant itself is common in the north and can be found in sandy soil, especially along lake shores and streams. It has pink flowers. The cooked roots taste like carrots, but are even more nourishing.

        2. Wooly lousewort (fig. 177) is a low plant with wooly spikes or rose-colored flowers. The sulphur yellow root is large, sweet, and edible either cooked or raw. It is found on dry tundra regions of North America.

        3. Bistort or knotweed (fig. 178) is also found on the tundra. It has white or pink flowers which form a slender spike. The elongated leaves are smooth edged and come out of the stem near the soil level.

POISONOUS

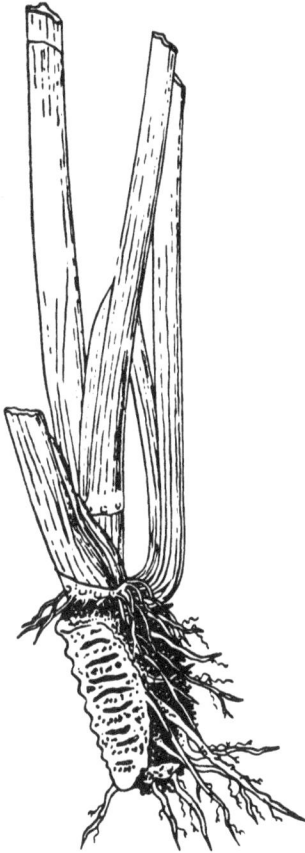

ROOTSTALK
WITH AIR CHAMBERS

Figure 169. Water hemlock.

The root is rich in starch but tastes slightly acid when eaten raw. It is best when soaked in water for a few hours and then roasted.

(d) *Antiscurvy plants.* These include scurvy grass (fig. 179) and spruce (fig. 180). Scurvy is a dread disease caused by the lack of vitamin C. It can be prevented by including the above plants, both rich in vitamin C, in your diet. All parts of scurvy grass are flaky,

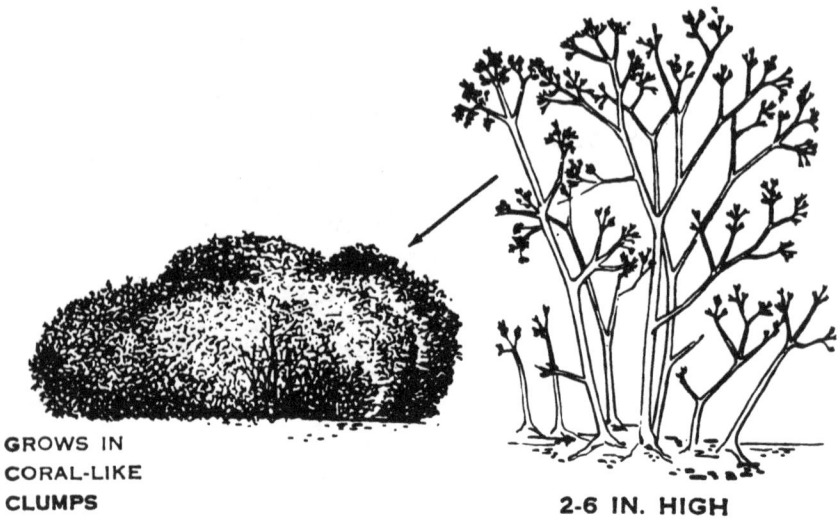

GROWS IN
CORAL-LIKE
CLUMPS

2-6 IN. HIGH

Figure 170. Reindeer moss (Lichen).

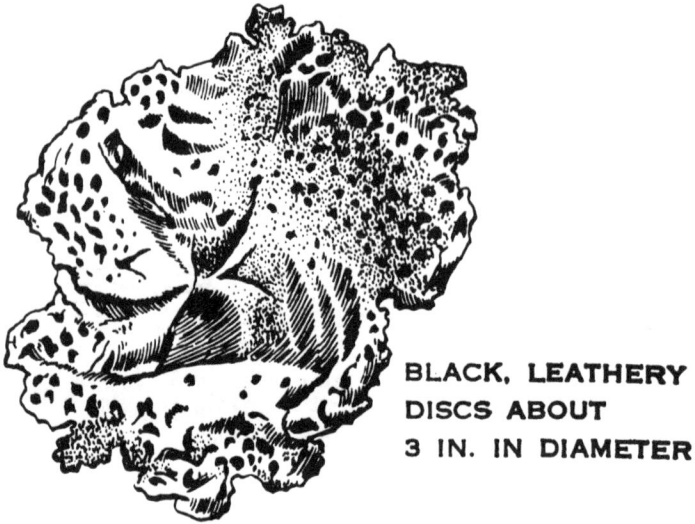

BLACK, LEATHERY
DISCS ABOUT
3 IN. IN DIAMETER

Figure 171. Rock tripe (Lichen).

and leaves, stems, and fruits can be eaten raw. Spruce twigs, leaves, bark, and stems are edible.

(e) *Greens*. Several polar plants are good substitutes for the leafy vegetables you eat regularly as part of your normal everyday life.

1. Wild rhubarb. See chapter 4.

2. Dandelion (fig. 181) is a potential life saver in polar regions. You can eat both leaves and roots raw, but

EDIBLE BERRY

Figure 172.   Salmonberry.

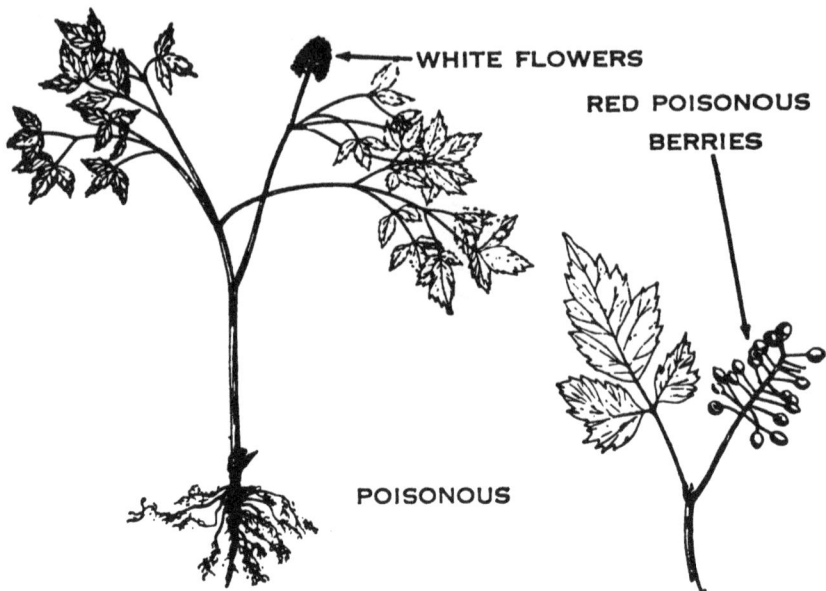

WHITE FLOWERS

RED POISONOUS
BERRIES

POISONOUS

Figure 173. Baneberry.

they taste better after light cooking. Dandelion roots can be used as a coffee substitute. To prepare, clean the roots, split, and cut them into small pieces. Roast them and grind the roasted roots between two stones. Cook like coffee.

3. Marsh marigold (fig. 182) is found in swamps and along streams and come out early during the spring. The leaves and stems, particularly of young plants, are tasty when cooked.

4. Seaweeds are good as a supplement to a fish diet. See chapter 4.

59. Firemaking

a. *Keep Your Matches Dry.* Always carry a supply of matches in a waterproof container. They are a necessity, especially where snow and ice add to the problem of securing tinder for a fire started by primitive methods. Use your matches sparingly, and use a candle to start a fire.

b. *Selecting a Site.* Select a site where you and your fire are protected from wind. Standing timber or brush makes a good windbreak in wooded areas, but in open country you will have to provide some form of protection. A row of snow blocks, the shelter of a ridge, or a scooped-out side of a snowdrift will serve as a windbreak on the ice pack. A circular wall of brush, cut and

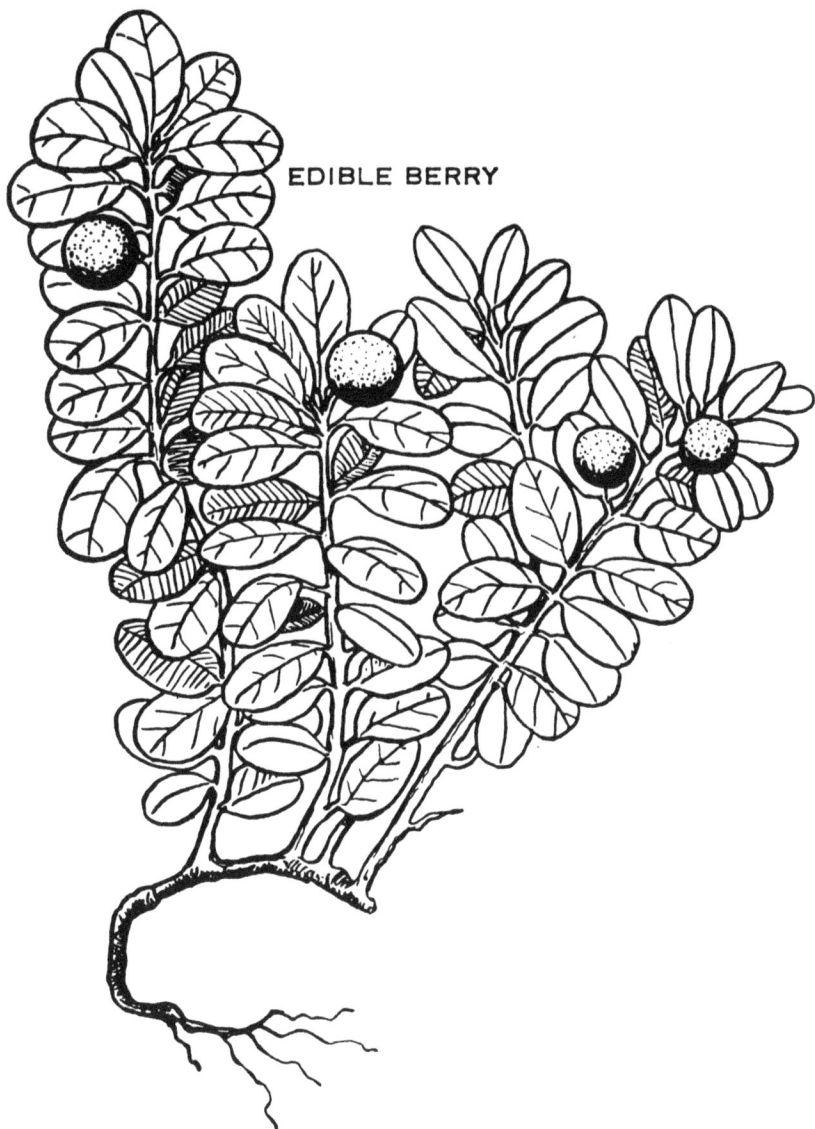

EDIBLE BERRY

Figure 174. Mountain berry.

stuck in the snow or ground, works well in willow country. A
ring of evergreen boughs is good in timber. Make your windbreak about four feet high and except for an entrance, let it
encircle the fire. Protect the fire from snow melting on overhanging tree limbs.

    c. Fuel.

        (1) Anything that burns is good fuel, and many kinds are
            available in the far north—animal blubber, lichens, ex-

EDIBLE BERRY

Figure 175. Alpine bearberry.

posed lumps of coal, driftwood, grass, and birch bark.
In some parts of the Arctic, however, your only fuel may
be animal fats, which you can burn in a metal container
by using a wick to ignite the fat. Seal blubber makes
a satisfactory fire without a container if gasoline or
heat tablets are available to provide an initial hot
flame. A square foot of blubber will burn for several
hours. Burned blubber cinders are edible.

(2) Fuel in subpolar regions is usually wood. The driest
wood is found in dead standing trees. In living trees,
branches above snow level are the driest. In the tundra
regions you can split green willows and birches into
fine pieces and burn them.

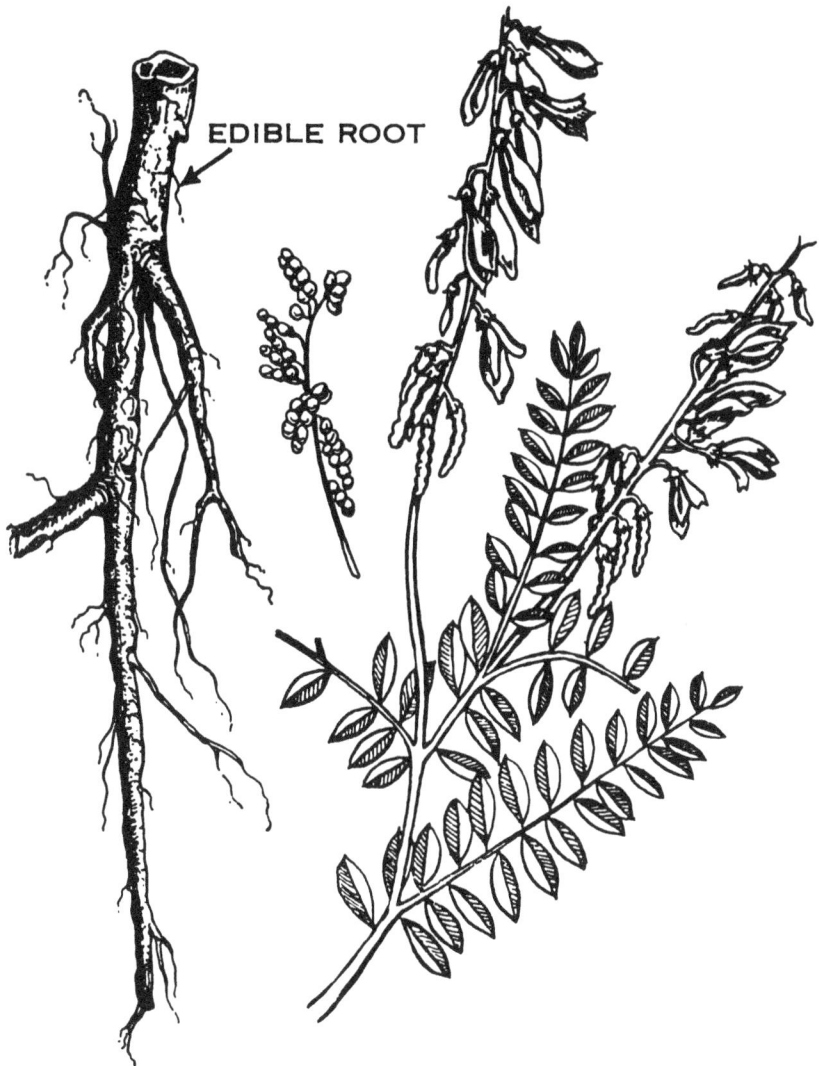

Figure 176. Sweet vetch.

(3) You can start a fire by primitive methods (ch. 5), but this is the last resort. If no matches are available, use gunpowder and cartridges. Construct a sheltered pile of kindling and wood. Place the powder from several cartridges at the base of the pile. Take two rocks and sprinkle a little powder on one rock. Then grind the two rocks together immediately above the powder at the base of the pile. This will ignite the powder on the rock and in turn, the larger powder.

EDIBLE ROOT

Figure 177.   Wooly lousewort.

EDIBLE ROOT

Figure 178.   Bistort.

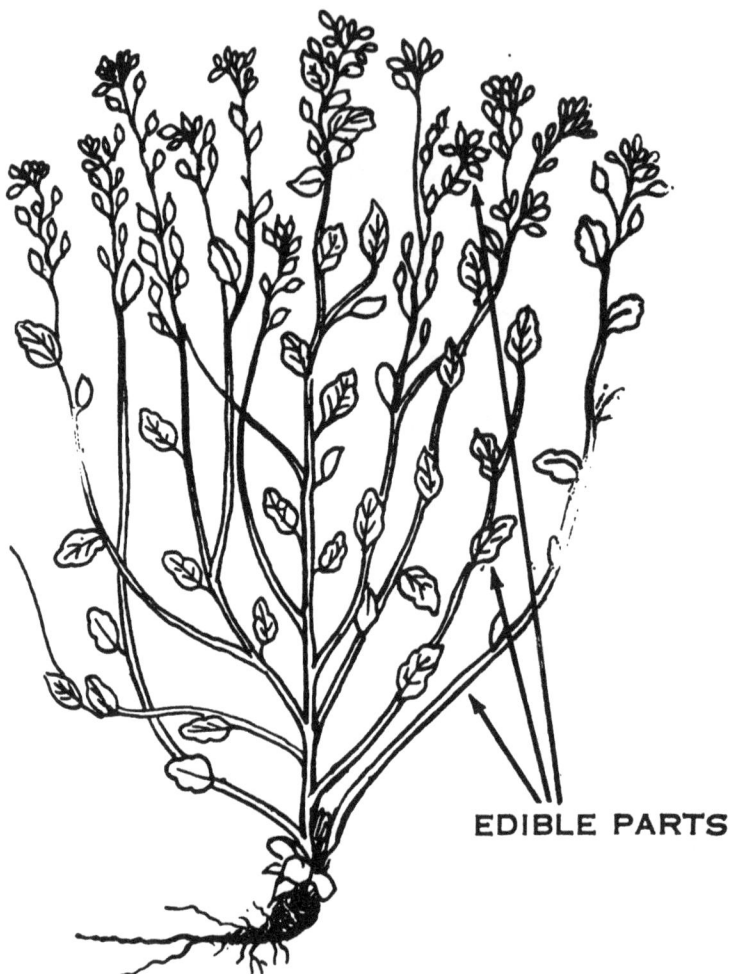

EDIBLE PARTS

Figure 179. Scurvy grass.

d. Cooking. Do not fry meat. This method eliminates the fat necessary for your health. See chapter 5.

60. Clothing

a. The basic survival problem in polar regions is **keeping warm.** Cold allows no time for trial and error experimentation. *You must do the right thing first.* Your clothing and how you wear it may determine how long you survive.

b. The body tends to maintain an average temperature of **98.6°.** In warm environments, the body absorbs the heat; in the cold, body heat is lost to the surrounding air. Clothes in cold climates then should serve one purpose—to keep body heat from escaping by insulating it against the cold outside air. Clothing of the

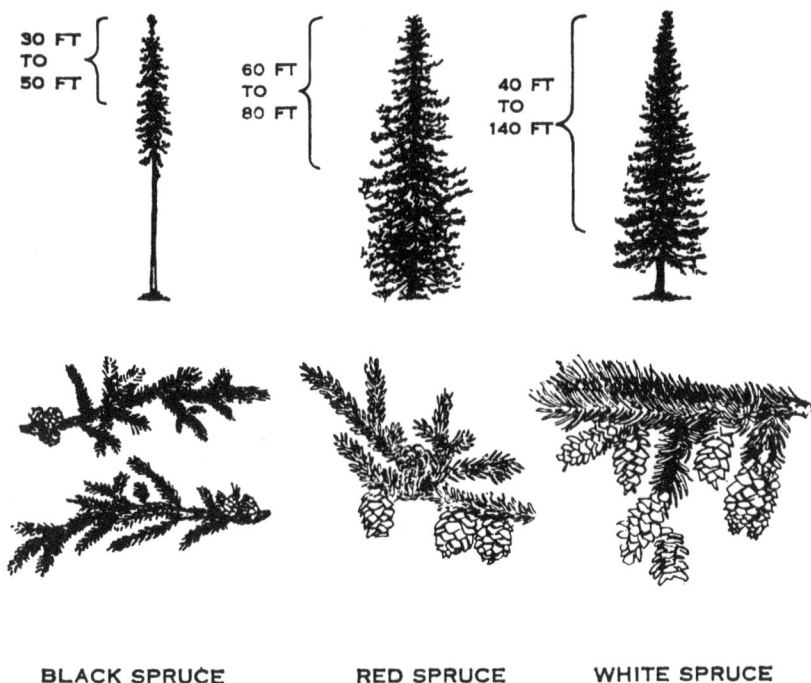

BLACK SPRUCE   RED SPRUCE   WHITE SPRUCE

Figure 180.   Spruce.

normal layer type, put on or removed as needed, helps to control body temperature. The inner layers of insulating clothing hold warm air in, while the wind resistant outer clothing keeps cold air from penetrating the clothing and carrying heat away.

    c. Some important facts about clothing and its relationship to you are—

        (1) Tight clothing reduces the zone of still air near your body and prevents free circulation of the blood.

        (2) Sweating is dangerous because it reduces the insulating value of clothing by replacing air with moisture. Also, as the sweat evaporates it cools your body. Avoid over-heating by removing layers of clothing and by opening your clothing at the neck, wrist, and front inclosure.

        (3) Body heat melts snow if you kneel or sit in one place for a length of time and will cause your clothing to become damp.

        (4) Hands and feet cool more quickly than other parts of the body and require special consideration. Keep your hands under cover as much as possible. They can be warmed by placing them next to the warm flesh under your armpits, between your thighs, or against your

ALL PARTS EDIBLE

Figure 181. Dandelion.

ribs. Feet, because they sweat more readily, are difficult to keep warm. However, you can be comfortable by wearing shoes large enough for you to wear at least two pairs of socks, and by keeping your feet dry. A warm double sock can be made by putting one pair of socks inside another and filling the space between them with a layer of dry grass, moss, or feathers (fig. 183).

(5) You frequently will have to improvise some articles of clothing like boots (fig. 184), especially if your boots are too small to allow for extra socks. A piece of canvas and some cord are all you need.

61. Health

a. Disease-transmitting insects, poisonous snakes, plants and animals, and diseases decrease as you move north or south from

EDIBLE LEAVES
AND STEMS

Figure 182. Marsh marigold.

the equator. Physical hazards such as snow and cold increase.
The chief danger to health in the Arctic is freezing. Snowblindness, carbon monoxide poisoning, and sunburn are secondary dangers. See chapter 1 for care of the feet.

  b. Frostbite is a constant danger to anyone exposed to temperatures below freezing. There is no particular pain to frostbite, and it can occur without you knowing it. The symptoms are
stiffness, lack of feeling, and greyish or whitish color of the
affected part. Warm the frostbitten area with a warm part of
your body, and do not exercise or massage the spot. *Do not apply*

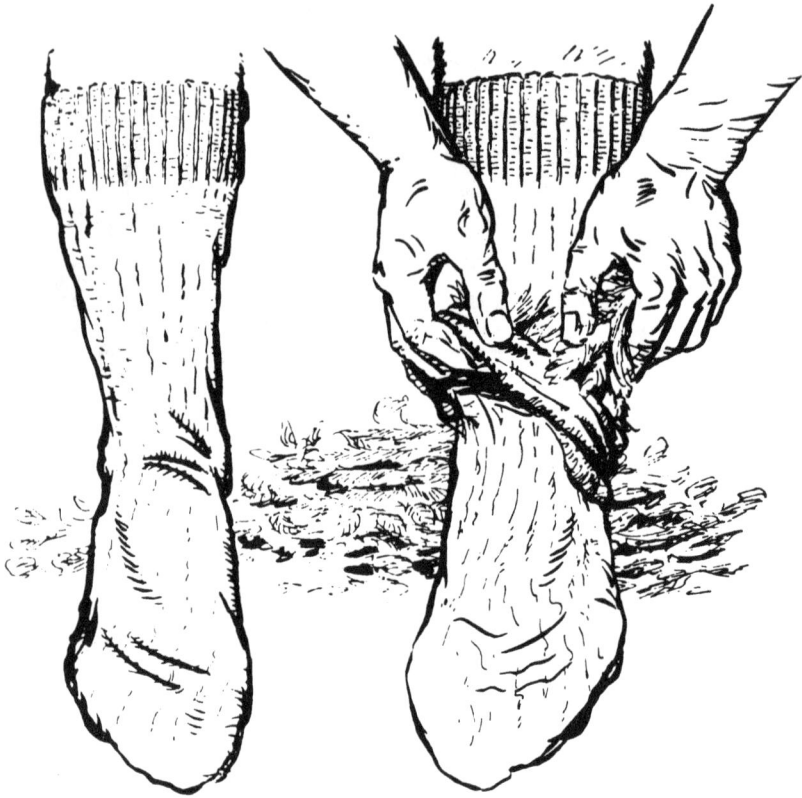

Figure 183. Insulating socks.

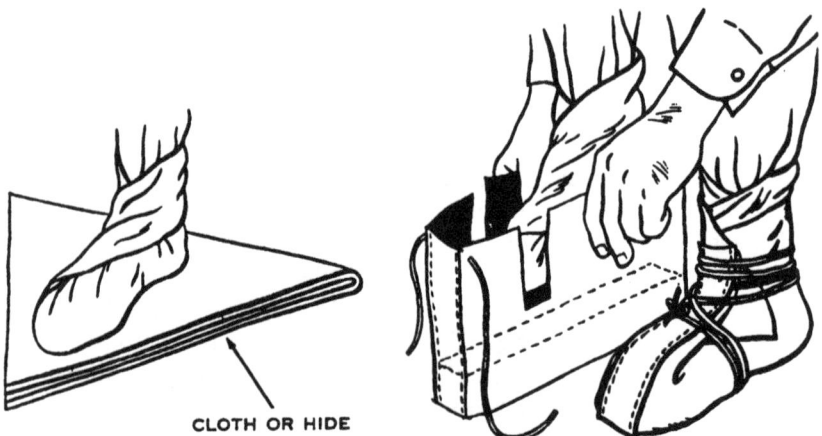

CLOTH OR HIDE

Figure 184. Improvised boots.

snow or ice. This treatment is dangerous. After frostbite there may be blistering and peeling just as in sunburn. Do not break or open blisters. Check exposed skin often, and if you have a companion check each other for symptoms of frostbite. To neglect frostbite is to invite gangrene.

c. Insufficient rest and improper diet greatly contribute to the risk of freezing to death. Symptoms of advanced general freezing are muscular weakness, fatigue, stiffness of limbs, and overpowering drowsiness. The victim's sight dims, he staggers, falls, and becomes unconscious. Give the victim something hot to drink. Treat him for shock. Rapidly rewarm any frozen parts of the body by immersion in warm water (90° F.–104° F.*); by placing a warm hand on the frozen part; or by exposure to warm air. Be especially careful in handling frozen body parts—frozen flesh can be easily damaged. Treat all cold injuries of the feet or legs as litter patients.

d. Snowblindness is caused by brilliant reflections or glare from the snow. It can occur even on foggy or cloudy days. The first warning of snowblindness comes when you no longer detect variations in the level of ground, followed by a burning sensation of the eyes. Later your eyes pain when exposed to even a weak light. Prevention is the best cure, but if you are stricken, complete darkness is your best medicine. Wear your sunglasses at all times. If none are available, use a piece of wood, leather, or other material with narrow eye slits cut in it (fig. 185). Glare is reduced if you blacken your nose and cheeks with soot.

e. (1) The danger of suffocation by carbon monoxide is a great hazard in the Arctic. To one subjected to extreme cold, the desire to get warm and stay warm often overrules common sense. Depend on your clothing to keep you warm—not a fire. In temporary shelters, use fires and heaters only for cooking. Any type fuel burning for as little as a half hour in a poorly ventilated shelter can produce a dangerous amount of odorless carbon monoxide fumes. Ventilation can be provided by leaving the top of the shelter open and another opening (for fresh air) close to the ground (door flap partially open) or by building a draft tunnel. The tunnel is buried in the floor and has an opening under the stove. The draft of the stove draws fresh air from the outside of the tent into the tunnel.

(2) If you are in a shelter and begin to feel drowsy, watch for carbon monoxide poisoning. Get into fresh air.

---

*With no thermometer, this temperature range may be estimated as being a temperature which does not feel too warm to the back side of an *undamaged* hand.

Figure 185. Improvised sunglasses

Move slowly and breathe evenly. *Above all, remove the source of the fumes.* If several men are sleeping in a closed, heated shelter, one man should stay awake to watch for indications of carbon monoxide.

f. A tourniquet should be applied only after all other means of control fail to stop bleeding. Once applied, the tourniquet must be left on, despite the probable loss of a limb due to freezing, since no replacement for lost blood will be available; it is better to lose a limb than to lose a life. Wounds which do not require tourniquet should be bandaged only tightly enough to stanch the bleeding, and loosened when bleeding has been controlled. Keep the body and affected limb or limbs comfortably warm at all times but do not overheat.

g. Arctic sunburn is possible on both cloudy and sunny days, and you should consider it a dangerous possibility. Prevention is the best cure. Animal tallow rubbed on the skin helps to prevent sunburn. A face mask similar to that used to prevent frostbite is also effective. A stubby beard also protects against sunburn. If you become sunburned, keep the affected parts moist with animal oils, and stay out of the sun.

h. (1) Just as in other areas, in the Arctic you must take good care of your body. Try to keep clean. If you cannot wash your body, at least try to keep your face, hands, armpits, crotch, and feet clean by wiping them with a cloth. Every night before you go to sleep, remove your boots, dry your feet, and rub and exercise them. Make provisions for drying your boots by holding them over your camp fire. Do not sleep in wet socks. Put them inside your shirt next to your body to dry. Wear two pairs of socks at a time if you have extra pairs and if your boots are large enough to permit it without interfering with circulation. If you have no fire and if your shoes are wet when you go to bed at night, stuff them with dry grass or moss in order to speed drying.

(2) Do not be afraid to relieve yourself. The areas of the body which you must expose will not stay exposed long enough to hurt you. Bury your garbage and your body wastes at a distance from your shelter and your water supply.

62. Natives

Natives are relatively few in the Arctic and those found in North America and Greenland are friendly. Eskimos live mostly along the coasts. Indians are found along rivers and streams of the interior. Arctic natives, like yourself, have little enough to eat, so don't take advantage of their hospitality. Offer payment when you leave (par. 52).

Section III. IN JUNGLE AND TROPICAL AREAS

63. Know Your Jungle

a. There is no standard jungle. It may be a tropical rain forest, which is the popular conception, or dry open scrub country. Jungle vegetation depends on climate and, to a large extent, the influence of man through the centuries. Tropical trees take over 100 years to mature and are fully grown only in untouched primeval virgin forest. This is a "primary" jungle and is easily recognized by its abundance of giant trees. The tops of these trees form a dense canopy over 100 feet from the ground, under which there is little light and underbrush. Land in this type jungle is difficult but not impossible to traverse.

b. Primary jungle growth has been cleared in many areas of the world to allow for cultivation. This land is later left idle and jungle growth reclaims it, making it a sea of dense underbrush and creepers. This is "secondary" jungle and is much harder to traverse than primary jungle.

c. Well over half the land in the tropics is cultivated in some way or other, and you will find rubber and tea plantations, coconut plantations, and native allotments. If you should find yourself in a plantation area, watch for the men who tend the crop—they may offer aid.

d. The tropical rain jungle, whether secondary or primary, is an unpleasant land to live in and travel through. The soil is covered with decaying vegetation over which countless millions of leeches move. Loathsome and bothersome insects abound. In secondary forest, sometimes impenetrable undergrowth bars travel. Over the undergrowth in primary jungle is the rather open space beneath the jungle tree tops where other trees, vines, and creepers grow. Birds and small animals are often found in this area. High among the tree tops of primary growth may be found birds, bees, moths, monkeys.

e. Dry scrub country is more open than the wet jungle but is difficult to travel because of its lack of topographical features, population, and tracks. It can be traversed, however, if you use a compass, patience, and common sense.

64. Travel

You can travel safely in the jungle if you don't become panicky. Alone in the jungle, depending on the circumstances, your first move is to sit down and relax and think your problem out. You should—

a. Pinpoint your position as accurately as possible to determine a general line of travel to safety. If you don't have a compass, use the sun in connection with a watch as an aid to direction (ch. 2).

b. Take stock of your water supply and rations.

c. Move in one direction but not in a straight line. Avoid obstacles; don't fight them. In enemy territory, take advantage of natural cover and concealment.

d. Turn your shoulders, shift your hips, bend your body, and shorten or lengthen, slow or speed your pace as required. There is a technique for moving through jungle; blundering only leads to bruises and scratches.

e. Develop a "jungle eye." See chapter 2.

f. Keep alert.

g. Check your bearings often.

h. Follow paths, trails, and boundaries if not over 20° off your axis of advance.

i. Travel at a steady pace.

j. Consider the advantages of traveling by stream (ch. 2).

65. Shelter

a. Selecting a Site.

(1) Try to pick a campsite on a knob or high spot of ground in an open place well back from swamps. You'll be bothered less by mosquitos, the ground will be drier, and there will be more chance for a breeze.

(2) Don't build under large trees with dead limbs. They may fall. Don't place your shelter under a coconut tree.

(3) In mountainous jungle the nights are cold. Get out of the wind.

(4) Avoid dry riverbeds—they can be flooded in a few hours, sometimes by rains so distant from you that you are not aware that any rain has fallen.

(5) If camping during the day and moving at night, the shady borders of rivers supply the best sites.

b. Types of Shelter. The type of shelter you build depends on the time available for preparing it and whether it is to serve as a permanent or semipermanent structure. Some jungle shelters are—

(1) Simple parachute shelter made by draping a parachute over a rope or vine stretched between two trees.

(2) Thatch shelter made by covering an A-type framework (fig. 186), with a good thickness of palm or other wood leaves, pieces of bark, or mats of grass. Slant the thatch shingle-fashion from the bottom upward. This type of shelter is considered ideal since it can be made completely waterproof. Use the broad leaves of young banana trees. Build a hot fire on a flat stone or a platform of small stones. When the stones are well heated place a leaf on them and allow it to turn dark and glossy. At this stage the leaf is more water repellent and durable and can be used as a shingle. After you finish your shelter, dig a small drainage ditch just outside its lanes and leading downhill; it will keep the floor dry.

c. Beds. Don't sleep on the ground; make yourself a bed of bamboo or small branches covered with palm leaves (fig. 187).

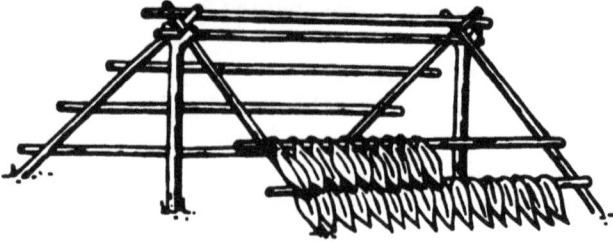

Figure 186. A-type frame work.

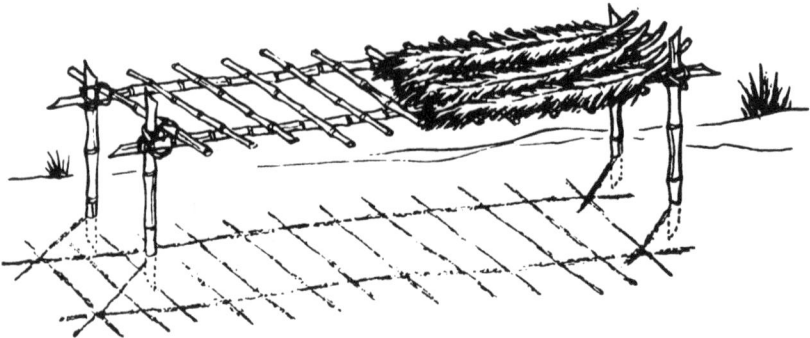

Figure 187. Bamboo bed.

A parachute hammock may serve the purpose. You can make a crude cover from tree branches or ferns; even the bark from a dead tree is better than nothing.

66. Water

a. Finding water in jungle country is usually not difficult. Use these hints—

    (1) The water from clear, swift-flowing streams containing boulders is a good source for drinking and bathing. But

before drinking it, purify by boiling or chemical means.

(2) You can get water that is almost clear from muddy streams or lakes by digging a hole 1 to 6 feet from the bank. Allow water to seep in and the mud to settle.

(3) Water from tropical streams, pools, springs, and swamps is safe to drink only after it has been purified.

(4) Collect rain water by digging a hole and lining it with tarpaulin or piece of canvas.

(5) Water can be obtained from vines and plants. Vines are good sources, (ch. 3). Bamboo stems often contain water.

(6) Animal trails often lead to water.

b. Coconuts, particularly when green, supply milk which is both pleasant and nourishing. A sugary sap can be obtained by cutting the flower spikes. Nuts are available throughout the year. A drinkable sugary sap can also be obtained from the buri, nipa, sugar, and other palm trees (ch. 3).

67. Food

a. *Abundance.* There is an abundance of food in the jungles, but some is poisonous (ch. 7). Any foods eaten by a monkey are generally safe for human consumption. If in doubt, eat the food in small quantities and find out what happens. It is wise to eat strange foods slowly and in small quantities. In inhabited areas of the tropics, practically all cultivated fruits and vegetables and/or foods which have been handled by natives have been contaminated with disease germs. Cultivated vegetables and fruits are usually fertilized with human wastes and are a source of infection. Never eat these fruits and vegetables raw unless you have peeled them or cut off all of the outer surface with a knife. To be really safe, cook all vegetables before you eat them. Be sure to protect your food from flies, because they can reinfect it after it has been cooked.

b. *Fish.* There are some poisonous fish in tropical waters (ch. 7), but there are many edible species also. The safest fish to eat are those from the open sea or deep water beyond the reef. The enterprising survivor can eke out a good living along the beach by eating shellfish, snails, snakes, lobsters, sea urchins, and small octopuses. Suckers abound in most tropical streams.

(1) There are no simple rules for telling whether a fish is undersirable as food. Often fish that are edible in one area may be unfit as food in another, depending on the place, their food, or even the season of the year. Eat only small portions of any fish. If you feel no ill effects, it is safe to continue eating (ch. 4).

(2) Tropical fish spoil quickly and should be eaten soon after they are caught. *Never eat entrails or eggs of any tropical fish.*

(3) Remember that fishing methods used at home are likely to prove successful in the jungle (ch. 4).

c. (1) Plants. See chapter 4.

(2) Some plants that are poisonous (ch. 7) and should be avoided are—

(a) White mangrove or blind-your-eye (fig. 188). This plant is found in mangrove swamps, estuaries, or on coasts. Sap causes blistering on contact. *It will blind if it contacts your eyes.*

(b) Cowhage or cowitch (fig. 189). This plant is usually found in thickets and brush country but never in true forest. The hairs of the flowers and pods cause irritation. Blindness results if they contact your eyes.

(c) Nettle trees (fig. 190). This plant is widespread, especially in and near ponds. It is poisonous to touch and causes a burning sensation.

(d) Thorn apple or jimson weed (fig. 191). This is a common weed of waste and cultivated land. All parts of this plant, especially the seeds, are poisonous.

(e) Pangi (fig. 192). This plant is found mainly in the Malayan jungle. Its seeds contain prussic acid.

(f) Physic nut (fig. 193). The seeds of the physic plant are violently purgative.

(g) Castor oil plant (fig. 194). A shrublike plant common in thickets and open sites, it has seeds which are poisonous and violently purgative.

68. Firemaking and Cooking

a. *Firemaking.*

(1) *Selecting a site.* Pick a dry, sheltered spot where there is least danger of the fire spreading. In the wet season a dry spot under a leaning tree may be available.

(2) *Kindling.* Good natural kindling materials are palm leaves, dry wood, bark, twigs, loose ground lichens, dead upright grass, and ferns. Shavings or wet, green bark from growing saplings are highly inflammable and can be used as kindling (ch. 5).

(3) *Fuel.* Anything that will burn will do—standing dead wood and dry dead branches, even some green wood. The hearts of dead logs burn readily. Look for dead wood hanging in overhead vines and branches. Green

Figure 188. White mangrove.

leaves thrown on a fire create a smudge that helps to keep off mosquitoes.

    (4) *Starting the fire.* Some of the methods discussed in chapter 5 may work in the jungle.

    b. Cooking (ch. 5). Jungle cooking may best be done by boiling and baking. Bamboo sticks are useful as liquid containers (ch. 5).

69. Clothing

    a. Unless completely covered, your body is vulnerable to leeches, insects, scratches, bruises, and cuts.

    b. You should have—

    (1) Clothing loose enough to be tucked into gloves and socks.

    (2) Clothing strong enough to withstand hard wear.

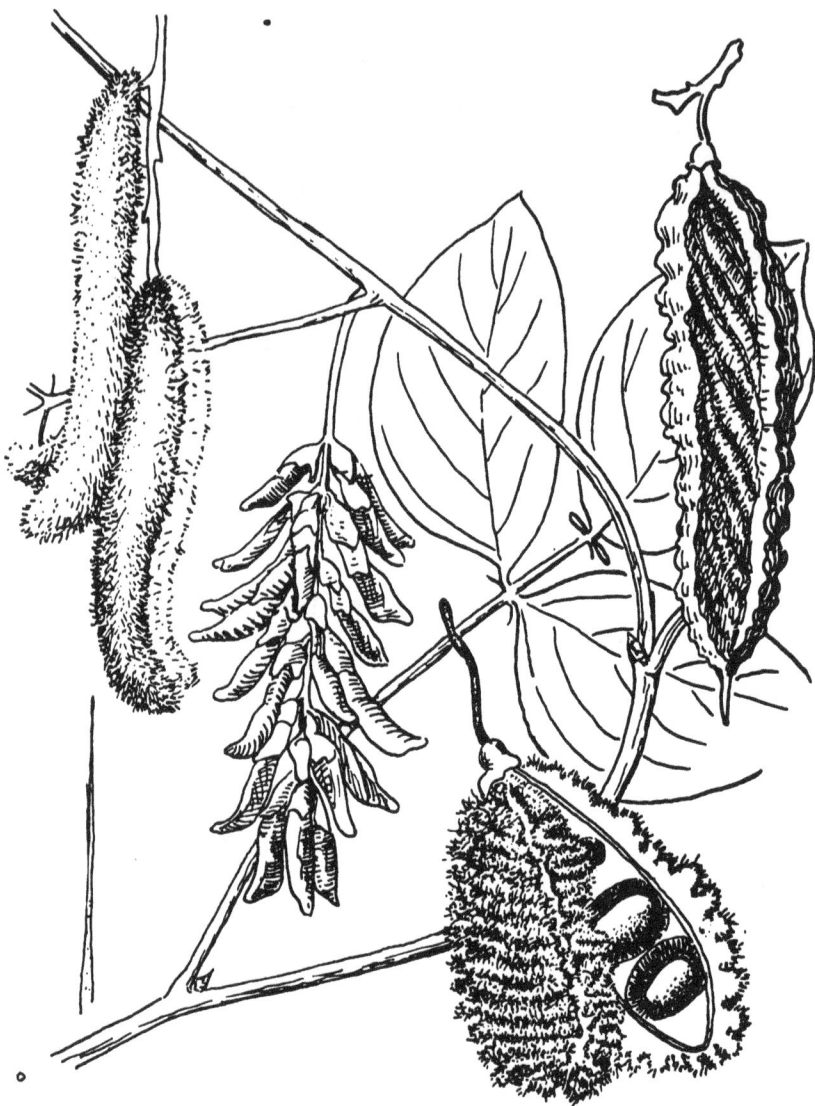

Figure 189.   Cowhage or cowitch.

   (3) Mosquito headnets and thorn-resisting gloves.
   (4) Plenty of pockets for carrying emergency items such as maps, compass, and matches.
   (5) Army boots. These are the best for jungle footwear.

70. Health

  a. General. You can't hope to elude the enemy and remain alive in the jungle areas unless you keep your body strong.

Figure 190.   Nettle tree.

Under ideal conditions this is often difficult, but your chances are increased by some commonsense rules.

(1) Don't hurry. Never try to beat the jungle by speed—you can't.

(2) Avoid climbing high terrain except for taking bearings. A long detour over flat ground is preferable.

(3) Take care of your feet by changing and washing your socks often. Also protect your footwear from cracking and rotting by using grease.

Figure 191.   Thorn apple.

(4) Should you get fever, make no attempt to travel. Wait until the fever abates. Drink plenty of water.

(5) Avoid ticks, leeches, mosquitoes, insects, and other pests (ch. 7). These constitute a real danger to your health and safety.

(6) Avoid infections. In the heat and dampness of the tropics, wounds quickly become infected. Try to protect a wound or sore by covering it with a clean dressing. Sterilize this bandage if possible.

(7) Prevent heat exhaustion, heat cramps, or heat stroke by replacing the water and salt you lose through sweating. Drink plenty of purified water; if you have salt, mix two tablets to a canteen (quart) of water. If you feel the effects of heat, relax in the shade and drink a

Figure 192. Pangi.

half canteen of this salted water every 15 minutes.
Continue this treatment until you feel better.

(8) Avoid sunburn.

b. Sickness and Fever. Diseases common to tropical areas
nclude—

(1) *Malaria* (ch. 7).

(2) *Dysentery.* Caused by polluted food or drinking water.
See chapter 3.

(3) *Sand fly fever.* This has symptoms similar to malaria.
Give plenty of water or other liquids by mouth and have
the patient rest until his fever abates. If you have
aspirins or "APC's," two tablets every four hours may
be given until six tablets have been taken.

(4) *Typhus.* There are several types of typhus to be found

Figure 193. Physic nut.

in tropical areas among which are the flea-borne, louse-
borne, and mite-borne varieties. The general symptoms
are severe headaches, weakness, fever, and generalized
body aches. The victims usually have a dusky complex-
ion and may or may not develop a pink mottled or
splotchy rash. Typhus of all types will likely prove
fatal without medical attention. Strict attention to
personal hygiene, avoidance of contact with lice or flea
ridden rodents, and avoidance of mite-infested grassy

Figure 194. Castor oil plant.

areas is essential. Inoculations for the louse-borne typhus infection are effective. Keep them current.

(5) *Dengue fever* (ch. 7).

(6) *Yellow fever* (ch. 7).

71. Natives

Ninety percent of the survival stories coming out of the Southwest Pacific during World War II stated that most jungle natives contacted were helpful. Where natives proved to be unfriendly or treacherous they did so because—

a. They were afraid of enemy occupation forces.

b. The survivors displayed weapons or threatened them with guns, knives, or clubs.

c. The survivors were too liberal in their display of money or in actual rewards of money or equipment.

Section IV. IN DESERT AREAS

72. Distribution

a. Those areas called "deserts" vary from salt deserts to rock deserts to sand deserts. Some are barren of plant and animal life; in others there are grasses and thorny bushes where camels goats, or even sheep can nibble enough to live. Anywhere you find them, however, deserts are places of extremes—extremely hot during the day, extremely cold at night, extremely free of plants or trees or lakes and rivers. Deserts are found throughout the world and comprise nearly one-fifth of the earth's surface (fig. 195).

b. Among the better known desert areas are the Sahara, Arabian, Gobi, and the flat plains of the southwest United States.

   (1) The Sahara is the largest desert in the world, stretching across North Africa from the Atlantic Ocean to the Red Sea, and from the Sahara Atlas Mountains in the north to the Niger River in tropical Africa. Most of the Sahara is gravel plain from which the sand has been blown away; but there are some desert mountains which rise 11,000 feet above sea level.

   (2) The Arabian desert covers most of the Arabian peninsula except for fertile fringes along the Mediterranean, the Red Sea, the Arabian Sea, and the valleys of the Tigris and the Euphrates Rivers. Since the Arabian desert is a source of oil, modern communities have been established along the desert's edges. The presence of this civilization has increased your chances for a safe journey afoot in this area. However, the desert of Arabia is treacherous and even native Arabs sometimes get lost.

   (3) The Gobi is one of the deserts of central Asia and is important because of its strategic location between Russia and China. The Gobi of Asia is not a starkly barren waste like the great African desert. Everywhere there is some grass, although it's often scanty. There are some distinct and well-channeled water courses leading from the surrounding mountains, but these courses are usually dry.

   (4) The deserts of the southwest United States generally

have more varied vegetation; a greater variety of scenery; and a more rugged landscape than either the Sahara or the Gobi. But like other deserts there is a marked decrease in moisture, and it is often a long time between drinks.

73. Travel

a. Water is the most important factor in desert survival. Carry all you can, even if you have to leave something else behind. When you decide to travel—

(1) Travel only in the evening or early morning. Stay in the shade and rest during the hot part of the day.

(2) Head for a coast, a known route of travel, a water source, or an inhabited area. Along a coast you can conserve your sweat by wetting your clothes in the sea.

(3) Follow the easiest route possible. This means you should avoid loose sand and rough terrain, and follow trails. In sand dune areas, follow the hard floor valleys between dunes, or travel on dune ridges.

(4) Avoid following streams hoping to reach the sea, except in coastal desert areas or those areas with large rivers flowing across them. In most deserts, valleys lead to an inclosed basin or temporary lake.

(5) Dress properly to protect yourself against direct sunlight and excessive evaporation of sweat. If you do not have sunglasses, make slit goggles (fig. 185). Remember that clothing is necessary for warmth on the desert because cool nights are common.

(6) Care for your feet. Your boots are perfect for desert travel. You can cross sand dunes barefooted in cool weather, but during the summer the sand will burn your feet. If you follow a caravan trail you will not encounter loose sand or broken, rocky areas.

(7) Check maps for accuracy if possible. Maps of desert regions are usually inaccurate.

(8) Take shelter during a sandstorm. Don't try to travel when visibility is bad. Mark your direction with a deep scratched arrow on the ground, a row of stones, or anything available. Lie down with your back to the wind and sleep out the storm. Cover your face with a cloth. Don't worry about being buried by the sand. Even in sand dune areas it takes years for the sand to cover a dead camel. If possible seek out some shelter on the lee of a hill.

(9) Multiply your estimation of distances by three since

the absence of features often makes an underestimate likely. See chapter 2.

b. Don't overlook the possibility of making yourself mobile. In both the Sahara and the Gobi, as well as in much of the American desert country, travel by ordinary automobile is practical. Native transportation, camels, horses, and donkeys may be available. The camel, although cantankerous and slow, can travel as long as 8 to 10 days without water.

74. Shelter

Shelter from sun and heat and the occasional sandstorm is your most important concern while surviving in desert areas. Since materials generally are not available for you to build a shelter, use these hints—

a. Get some protection from the sun by covering your body with sand. Burrowing in the sand also reduces water loss. Some desert survivors report that the pressure of the sand offers valuable physical relief to tired muscles.

b. If you have a parachute or other suitable cloth, dig out a depression and cover it. In rocky desert areas or where desert shrub, thorn shrub, or tufted grass hummocks grow, drape a parachute or blanket over the rocks or shrubs.

c. Make use of natural desert features for shade or shelter— a tree, a rock cairn, caves. The wall of a dry stream bed may provide shelter, but after a cloudburst your home may become suddenly flooded. Wadi-banks—along dried riverbeds, valleys, and ravines—are particularly good places to look for caves.

d. Utilize native shelter when practicable. Survivors reported during World War II that even desert tombs were used for protection against the elements.

75. Water

a. General. The importance of water cannot be overemphasized. You must have it, regardless of how adequate your food supply may be. In hot deserts you need a minimum of one gallon per day. If you control your sweating and travel during the cool desert night, you can move about 20 miles on that gallon. During the heat of the day you are lucky to travel 10 miles. Your time of expected survival under different conditions is shown in figure 196. The figures in columns 2–7 represent survival time in days.

b. Conserve Water.
  (1) Keep fully clothed. Clothing helps control sweating by not letting perspiration evaporate so fast that you miss some of its cooling effect. You may feel cooler

NO WALKING AT ALL . . .

MAX. DAILY IN SHADE TEMPERATURE (°F)	AVAILABLE WATER PER MAN, U.S. QUARTS					
	0	1	2	4	10	20
120	2	2	2	2.5	3	4.5
110	3	3	3.5	4	5	7
100	5	5.5	6	7	9.5	13.5
90	7	8	9	10.5	15	23
80	9	10	11	13	19	29
70	10	11	12	14	20.5	32
60	10	11	12	14	21	32
50	10	11	12	14.5	21	32

MAX. DAILY IN SHADE TEMPERATURE (°F)	AVAILABLE WATER PER MAN, U.S. QUARTS					
	0	1	2	4	10	20
120	1	2	2	2.5	3	
110	2	2	2.5	3	3.5	
100	3	3.5	3.5	4.5	5.5	
90	5	5.5	5.5	6.5	8	
80	7	7.5	8	9.5	11.5	
70	7.5	8	9	10.5	13.5	
60	8	8.5	9	11	14	
50	8	8.5	9	11	14	

WALKING AT NIGHT UNTIL EXHAUSTED AND RESTING THEREAFTER

Figure 196. Survival time.[17]

without your shirt, but you sweat more. And you might get sunburned.

(2) Stay in the shade. If possible sit up a few inches off the ground. The temperature is quite a few degrees lower off the sand.

(3) Don't hurry. You'll last longer on less water if you keep down your sweat. However, if you have sufficient water hurry all you need to.

(4) Don't use water for washing unless you have a sure and lasting supply.

---

[17]Adolph and Associates, Physiology and Man, New York, Interscience Publishers, Inc., 1947.

(5) Don't gulp your water. Drink in small sips. Use water only to moisten your lips if the supply is critical.

(6) Use salt only with water and only if you have an ample supply of water. Salt causes you to drink more.

(7) Keep small pebbles in your mouth or chew grass as a means of allaying thirst. Prevent water loss by breathing through your nose. Don't talk.

(8) Rationing yourself to 1 or 2 quarts of water a day is inviting disaster (at high temperatures) as such small amounts do not prevent dehydration (par. 79). Ration your sweat, not your water.

c. Locate Wells. A minimum of four quarts of water per day may be difficult to find unless you are near a well or oasis. Since wells are the source of most water on the desert, your best bet is to locate them as you travel along a native trail. There are other ways of locating water in the desert. Use these guides—

(1) Along sandy beaches or desert lakes, dig a hole in the first depression behind the first sand dune. Rain water from local showers collects there. Stop digging when you hit damp sand and allow the water to seep in. If you dig deeper you may hit salt water. (fig. 197).

(2) Scoop out a shallow well anywhere you find damp sand.

(3) Dry stream beds often have water just below the surface. It sinks at the lowest point on the outside of a bend in the channel as the stream dries up. Dig along these bends for water.

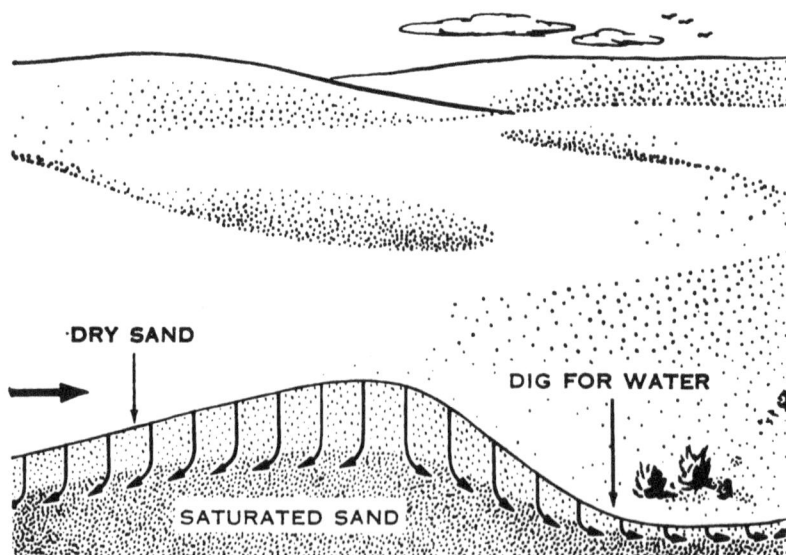

DRY SAND

DIG FOR WATER

SATURATED SAND

Figure 197. Locating water in sand dunes.

(4) Dew might be a source of water, particularly in some regions. Cool stones or any exposed metal surface will serve as a dew condenser. Wipe off the dew with a piece of cloth and wring it out. Remember that dew evaporates soon after sunrise and should be collected before then.

(5) Look for natural tanks or cisterns that may be located behind rocks, in gullies or side canyons, and under cliff edges. Often the ground near them is solid rock or hard-packed soil on which native paths do not show up. In the absence of such markers, search for the water point by observing animal droppings.

(6) Watch the flight of birds, particularly at sunset and dawn. Birds circle water holes in true desert areas.

(7) In the Gobi desert don't depend on plants as a source of water. The wild desert gourd may be considered on the Sahara (ch. 4). The large barrel cactus of the American desert also contains considerable moisture which can be squeezed out of the pulp. This is sometimes a difficult task, however, and your best bet is a well or another source. See chapter 3.

(8) Disregard the romantic stories of poisoned wells. These stories generally originate because of bad tasting water that contains salt, alkali, or magnesium.

(9) Purify all your water. This is especially true in native villages and around civilization.

76. Food

a. General. Food is difficult to find in the desert. However, food is secondary to water and you may get along without it for several days without any ill effects. Ration your food from the beginning. Eat nothing the first 24 hours and don't eat unless you have water.

b. Natural Sources.

(1) Animals are rare in the desert. You may find rats and lizards near a water hole and they may be your only diet. Deerlike animals are sometimes found in open desert country, but they are difficult to approach. The most common desert animals are small rodents (rabbits, prairie dogs, rats), snakes, and lizards, which are usually found near brush or water. Look for land snails on rocks and bushes.

(2) Some birds are found on the desert. Try kissing the back of your hand with a sucking sound to attract them to you. Sand grouse, bustard, pelicans, and even

gulls have been seen over some desert lakes. Use a baited deadfall or a hook and gorge to catch them (ch. 4).

(3) Usually where you find water you will find plants. Many desert plants look dry and unappetizing, so look for some soft part that is edible. Try all soft parts above the ground—flowers, fruits, seeds, young shoots, and bark. During certain seasons you may find some grass seeds or bean bushes. These beans grow on acacia trees that are often thorny and similar to mesquite or cat-claw of the southwestern United States (fig. 198). The prickly pear (a type of cactus) is a native of the North and South Americas and is widely found in North Africa, the near East, and Australian deserts. See chapter 4.

(4) All grasses are edible, but the ones found in the Sahara or Gobi are neither palatable nor nutritious. Try any plant you find. At least taste it—a taste will not kill you, even if the plant is poisonous. Dates may be available in the northern half of Africa, southwest Asia, and some parts of India and China.

c. *Native Food.*

(1) Native food in the Sahara is both palatable and edible. You are less likely to enjoy the food that is available in the Gobi since the native Mongols have less idea of cleanliness than the Arabs. To be safe, cook all of your

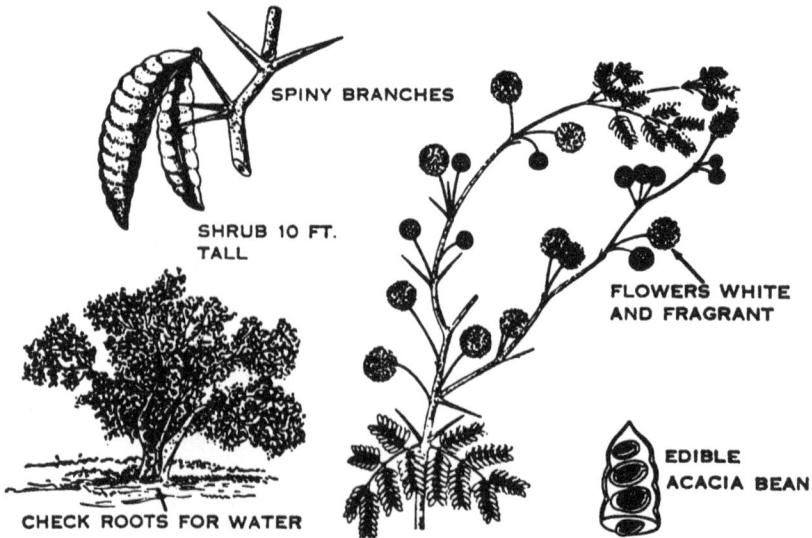

Figure 198. Acacia.

food. Appeal to the natural hospitality of the natives for food—don't steal it.

(2) Native dairy products—milk, cream, ice cream, butter, and cheese—are very dangerous. So are fruits and other cooked foods given to you by natives. If possible, trade or buy raw foods and prepare them yourself.

77. Firemaking and Cooking
(ch. 5)

a. Palm leaves and similar fuel are found in or near an oasis. But out on the open desert you must use any bit of dead vegetation you can find. Dried camel dung is used when there is no wood available.

b. Probably your most effective way of building a fire without matches will be sun rays directed through a magnifying glass. Other primitive methods of building a fire may not be available (ch. 5).

78. Clothing

a. You must protect yourself against direct sunlight, excessive evaporation of sweat, and the many annoying desert insects.

(1) Keep your body and head well covered during the day. Wear long pants and a shirt with long sleeves.

(2) Wear a cloth neckpiece hanging down to protect the back of your neck from the sun (fig. 199).

(3) If you must throw away some clothing in order to lighten your load, keep enough to protect you against the cold desert night.

(4) Wear clothing loosely.

(5) Open your clothing only when you are well shaded. Reflected sunlight can cause sunburn.

b. Protection of your feet may be the difference between life and death. The following hints will prove helpful:

(1) Keep sand and insects out of your shoes and socks, even if you have to stop often to clean them out.

(2) If you don't have boots, make some spiral puttees out of any available cloth. To do this cut two strips, each 3 or 4 inches wide and about 4 feet long. Wrap them spiral fashion around the top of your shoes and upward. This will keep out most of the sand.

(3) You can improvise a pair of sandals out of a sidewall of an old tire, if there are salvaged vehicles available (fig. 200). However, it is better to reinforce the soles of your shoes with heavy cloth if they are just worn.

(4) Remove your shoes and socks while you are resting in the shade. Use caution when doing this because your

Figure 199.   Cloth neck protector.

feet may swell, making it difficult for you to put your
shoes back on.

 (5) Don't try to walk barefooted. The sand will blister
your feet. Also, a barefoot hike across a salt flat or
mire will result in alkali burns.

 (6) Improvise clods to protect your feet around camp. Nail
a strap to pieces of wood and attach them to your feet.
Be sure to protect the tops of your feet from the sun.

79. Health

 a. Physical Hazards. All in all, the desert is a healthy place.
The heat and the effects of the sun coupled with a shortage or
complete lack of water are your most dangerous threats. Desert
sores, which develop from minor abrasions and which are slow
in healing, also may cause you some discomfort. Malaria does
not exist on the desert except in some oases of the Sahara. Your

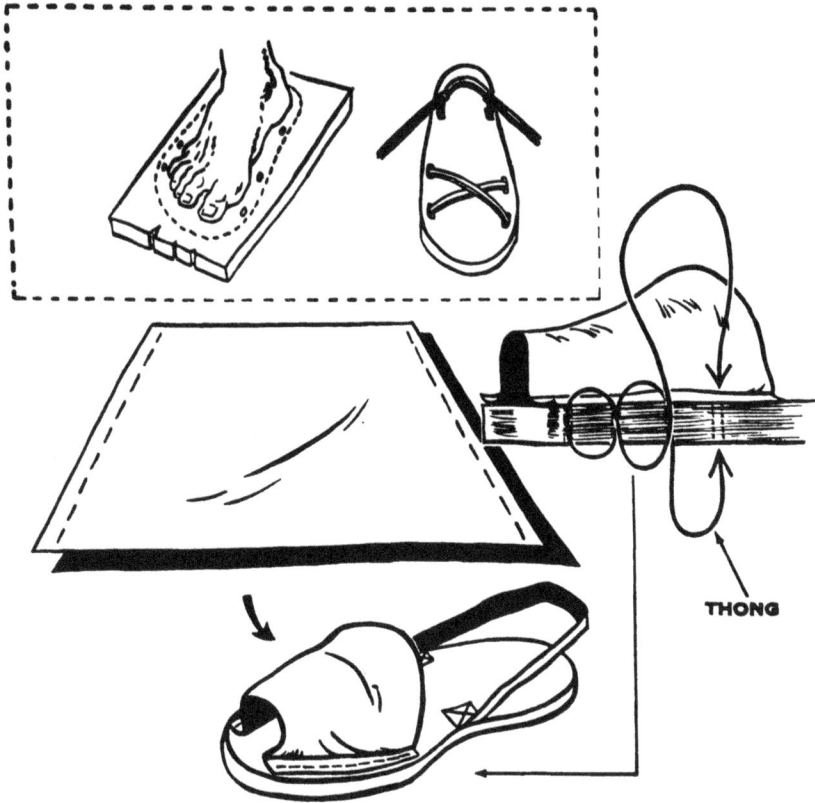

Figure 200. Improvised sandals.

inoculations will help to prevent typhoid or paratyphoid. You can avoid dysentery by watching your diet and not eating uncooked native food. Venereal diseases are prevalent, however. Probably you will remain free of disease if you follow your normal habits of cleanliness and diet.

b. Insects, Snakes, and Scorpions. Flies, sand fleas, and mosquitoes may prove a definite mental and physical hazard on the desert. Other insects are less numerous. In some regions you may encounter scorpions or snakes (ch. 7).

c. Dehydration. When your body loses water through sweating, it is called dehydration, and the water loss must be replaced by drinking. Otherwise you lose in efficiency. First you feel thirsty and uncomfortable; then you take it slow and have no appetite. As you lose more water, you get sleepy, your temperature goes up, and by the time you dehydrate as much as 5 percent of your body weight, you get nauseated. When temperatures are in the nineties and higher, 15 percent dehydration is probably fatal.

This may be as high as 25 percent in air temperatures of 85° F. or cooler. Luckily water returns you to normal if taken in time. Water in insufficient quantities does not prevent dehydration, but it does slow up the process.

80. Natives

a. Natives who walk the desert trails are friendly people, and like people living close to nature anywhere in the world, their natural sense of hospitality is strong. This is especially true if you appear friendly. A good rule to remember is the old adage, "When in Rome, do as the Romans do." Here are a few of the more important don'ts—

(1) Don't bawl out an offender in front of other people.
(2) Don't ask about a man's wife.
(3) Don't throw a coin at a man's feet—that is insulting.
(4) Don't swear at a native.
(5) Don't draw sand pictures or maps with your foot. Stoop down and draw it with your right hand.
(6) Don't gamble; it is forbidden.
(7) Don't expose the soles of your feet. Sit tailor fashion or on your heels.
(8) Don't be impatient.
(9) Don't act unfriendly.
(10) Don't fool around native women. Most tribes prescribe swift, sure punishment for violations of the code. When women are given liberty there is almost a veneral disease certainty.

b. Check with your commander for more specific information on the customs and forms of etiquette of the people in your area.

Section V. AT SEA

81. It Could Happen

a. General. There are many reasons why you may find yourself faced with the problem of surviving at sea. The ship or aircraft which you are aboard may be sunk or downed by natural hazards such as fire or collision, or by enemy action. Why you are adrift at sea is unimportant. The important question is how you will continue to remain alive. The answer depends to a great extent on the rations and equipment you have with you, the use you make of them, and your own skill and ingenuity.

b. Rations and Equipment. Lifeboats, rafts, and aircraft contain equipment adequate for emergencies at sea. You should know what this equipment is and where it is stored. Learn how to use it. Take care to see that fishing tackle is included. Fish may be.

your only source of food and water. See FM 21-22 for a lifeboat emergency equipment list, abandon-ship procedure, and command procedure aboard lifesaving craft.

82. Living Without Food and Water

a. Without water you will become delirious in about 4 days and die in 8 to 12 days. If you have water and no food you can expect to survive as long as 3 weeks. The reason you can live without food for longer periods is because of the body's ability to convert body fat and proteins into heat and energy, a pound of body fat providing the equivalent of two good meals.

b. Since your body uses up water in digesting and assimilating food and eliminating waste products, don't eat unless you have water and don't eat or drink for the first 24 hours. Then ration yourself. To conserve your body water, control sweating by protecting yourself from the sun and wind, avoiding exercise and remaining calm. Wet your clothing periodically to avoid sunburn.

83. Water

a. Rain, ice and the body fluids of animal life are your only natural sources of water at sea. *Sea water is not drinkable.* It aggravates your thirst and increases water loss by drawing body fluids out of the tissues to then be executed by way of the kidneys and intestines. To drink it is to invite delirium and death.

    (1) *Rain water.* Use buckets, cups, tin cans, sea anchor, boat cover, sails, strips of clean clothing, and all canvas gear in the boat to collect rain water. Devise your catchment before there is an actual need. If the shower promises to be light, wet your receptable in the sea. The salt contaminating the rain water will be slight, and the dampened cloth will prevent the fresh water from becoming absorbed in the fabric. Remember that your body can store water; therefore, drink all you can hold.

    (2) *Ice.* Sea ice loses its salt after a year and becomes a good source of water. This "old" ice is identified by rounded corners and its bluish color.

    (3) *Sea water.* In freezing weather, fresh water can be obtained from sea water. Collect this water in a container and allow it to freeze. Since the fresh water freezes first, the salt concentrates as a slush in the core of the frozen fresh water. Throw away this salt and the remaining ice is salt free enough to keep you alive.

b. Chemical kits may be available in your raft or lifeboat. These kits can be used to remove the salt and alkaline from sea water. Directions accompany the kits.

c. Observing water discipline.

(1) Drink slowly. After a rain drink your fill, but take an hour or more to do so. This prevents your kidneys from wasting water.

(2) Do not drink compass liquid. It is poisonous.

(3) Do not drink urine. It is dangerous.

(4) When water is scarce use it only to moisten your lips and throat. Gargle the water before swallowing it.

(5) Distinguish between true and artificial thirst. True thirst creates a burning irritation and complete dryness of the mouth and throat. Artificial thirst is brought on by the thought of water or its needs, or by eating foods or drinking water that is salty.

(6) Suck a button or chew gum to help relieve artificial thirst.

(7) Drink all the water you can before leaving the ship. you should go 24 hours before requiring another drink.

(8) Do not waste drinking water that has been contaminated with sea water. It can be used in small quantities with fresh water.

84. Food

a. General. The sea is rich in different forms of life. Your problem is to tap this source of food. If you have fishing equipment, chances are excellent that you will have food; but even if you have no equipment the situation is not hopeless.

b. Fish.

(1) *General.* Practically all freshly caught sea fish are palatable and wholesome, cooked or raw. In warm regions gut and bleed fish immediately after you catch them. Cut fish that you do not eat immediately into thin, narrow strips, and hang them to dry. A well-dried fish stays good for several days. Fish not cleaned and dried may spoil in half a day. Never eat a fish that has pale, shiny gills, sunken eyes, flabby skin and flesh, or unpleasant odor. Good fish should show the opposite characteristics. Sea fish should have a salt water or clean fishy odor. Eels are fish and good to eat, but do not confuse them with sea snakes. See chapter 7. The heart, blood, intestinal wall, and liver of fish are edible. Intestines should be cooked, however. The stomachs of large fish may contain partly digested smaller fish. These are edible. Sea turtles also are good food.

(2) *Fishing line.* You can make a strong fish line from pieces of tarpaulin or canvas by raveling the threads

and tying groups of three or more together in very short lengths. You can also use parachute shroud lines, shoelaces, or thread from clothing.

(3) *Fish hooks.* No one at sea should ever be without fishing equipment on his person, but even without fishing tackle you can improvise enough to survive.

    (a) Hooks may be made from items with points or pins, such as nail files, collar insignia and campaign ribbons; or, from bird bones, fish spines, and pieces of wood (fig. 201). See chapter 4. To make a wood hook, shape the shaft and cut a notch near the end to hold the point. Sharpen the point so the hardest part of the grain forms the tip of the hook. Use strands of canvas to lash the barb and shaft together.

    (b) Improvise fish lures by using a coin or snelled hook, or a dime fastened to a double hook (fig. 202).

(4) *Bait grapple.* You can collect and bring in seaweed by using an improvised grapple made of wood cut from your raft or boat. Use the heaviest piece of wood as the shaft, and cut three notches in which to fit three grap-

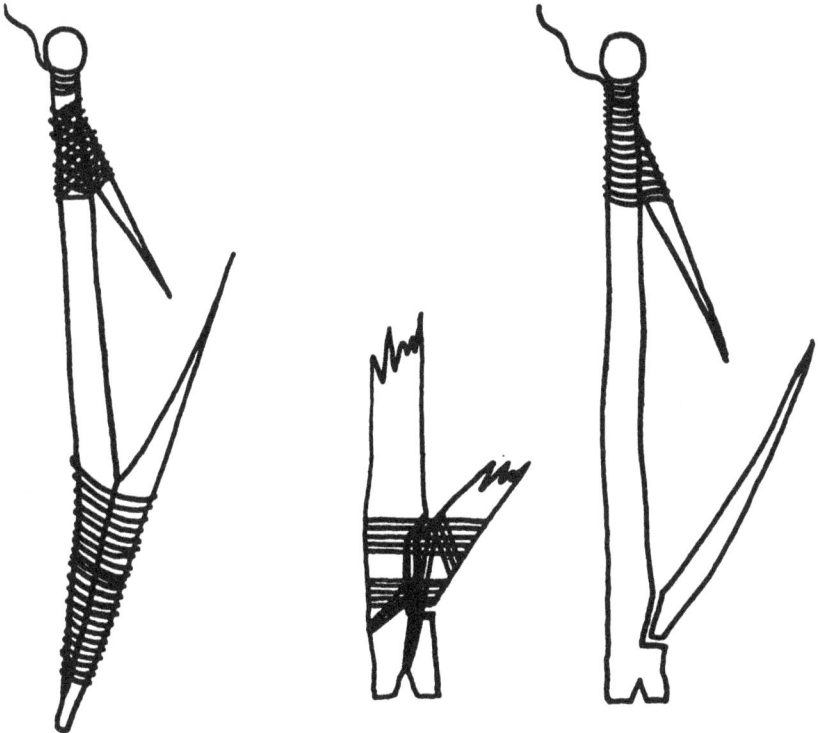

Figure 201. Fish hook from pieces of wood.

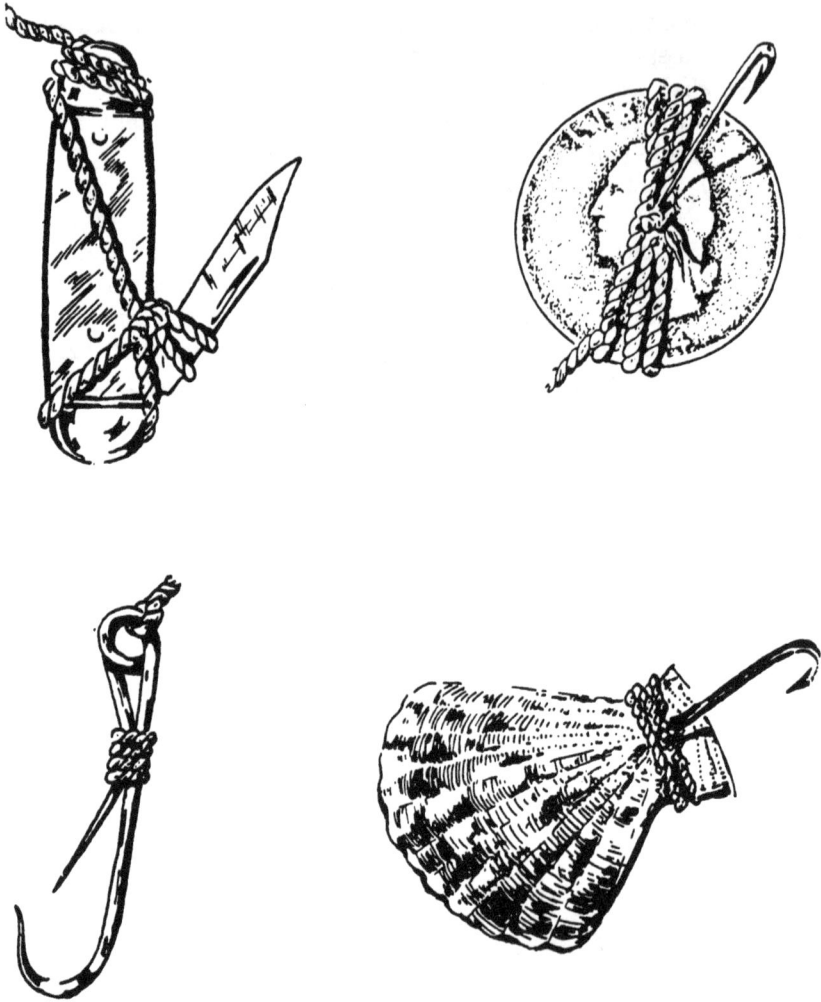

Figure 202. Artificial lures from coins and attached hooks.

ples. Lash them in place at a 45° angle (fig. 203). Tie a line to the shaft and drag the grapple behind your raft.

(5) *Bait.* Use small creatures found in the sea as bait for larger ones. Use the dip net from your fishing kit to scoop up these small creatures. If you have no kit, make a net from mosquito headnet, parachute cloth, or clothing fastened to oar sections. Hold the net under water and scoop upward. Save guts of birds, and fish for bait. Try a piece of colored cloth, bright tin, or even a button from your shirt. Try anything. Keep it moving

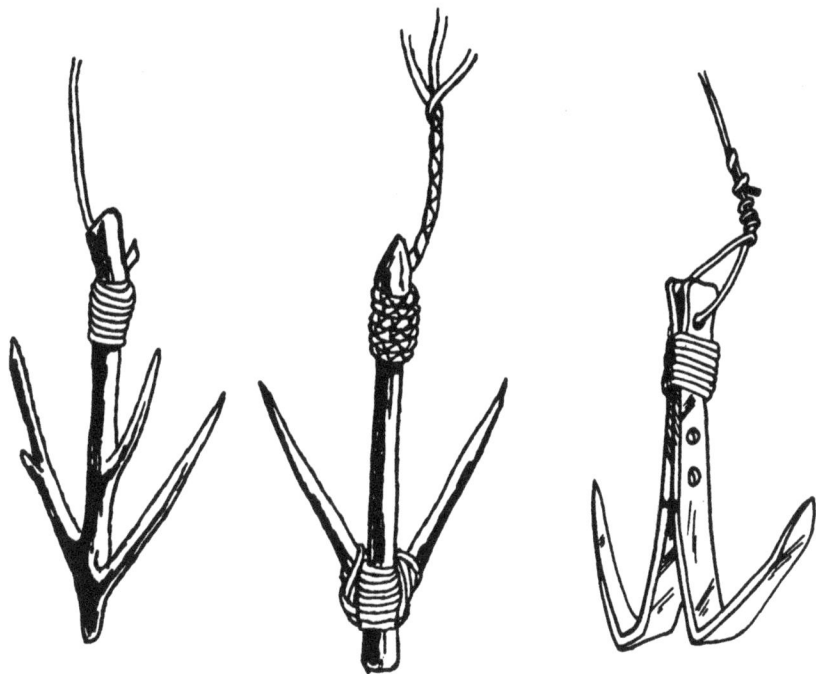

Figure 203.   Improvised grapples.

in the water to make it appear alive. Try it at different depths.

(6) *Fishing at sea.* Use the following hints while fishing at sea:

(a) Avoid spiny fish and those with teeth.

(b) Don't attach your line to anything solid; a big fish might break it. Don't wind the line around any part of your body.

(c) If you hook a large fish, take care not to capsize your raft or boat.

(d) In a rubber raft, be careful not to puncture it with hooks, knives, or spears.

(e) Fish for smaller fish. If sharks are near, don't fish.

(f) Watch for schools of fish which can be seen breaking water. Get close to a school if possible.

(g) Shine a flashlight on the water at night, or use a piece of canvas or cloth to reflect moonlight. The light will attract fish which may leap into your raft.

(h) Shade attracts many varieties of small fish, so lower your sail or tarpaulin into water. It may gather fish.

(i) Remember that the flesh of all fish caught in the open sea (except jellyfish and the liver of some fish) is good to eat, cooked or raw. Raw fish is neither salty nor unpleasant. See chapter 7 for list of poisonous fish.

(j) Make a spear or harpoon by tying a knife to an oar. Use this for large fish.

(k) If you lose your fishing equipment, try dangling a piece of fish or bird gut in the water. One survivor reported that he caught 80 fish in one day by allowing them to swallow a piece of gut and snatching them into his raft.

(l) Care for your equipment. Allow the lines to dry, and make sure your hooks are not sticking into the line. Clean your hooks.

c. *Seaweed.* Raw seaweeds are tough and salty and are difficult to digest. They absorb your body water, so eat them only if you have plenty of drinking water. However, seaweeds are an important survival item because they usually harbor small edible crabs, shrimp, and fish. Use a grapple to gather seaweed (fig. 203). Shake seaweed over your raft to reveal the small edibles.

d. *Birds.*

(1) Eat any bird you are fortunate enough to catch. Birds sometimes settle on the raft or boat, and survivors have

reported instances where birds landed on their shoulders. If birds are shy, try dragging a baited hook or throwing a baited hook in the air.

(2) Birds are relatively few on the North Atlantic, and are found mostly along the coasts. This is especially true in the North Pacific. In southern waters many species of birds often are seen hundreds of miles from land.

(3) Gulls, terns, gannets, and albatross can be caught by dragging a baited hook, or attracted within shooting distance by a bright piece of metal or shell dragged behind the raft. It is possible to catch a bird if it lands within your reach. However, most birds are shy and will settle on the raft out of reach. In this case, try a bird noose. Make it by tying a loose knot with two pieces of line as shown in figure 204. Bait the center of the hoop with fish entrails or similar bait. When the bird settles, tighten the noose around its feet. Use all parts of the bird; even its feathers can be stuffed inside your shirt or shoes for warmth.

85. Signs of Land

a. *Indications of Clouds*. Clouds and certain distinctive reflections in the sky are your most reliable indications of land. Small clouds hang over atolls and may hover over coral patches and hidden reefs. Fixed clouds or cloud crests often appear around the summits of hilly islands or coastal land. They are recognized easily because moving clouds pass by them. Other aerial indications of land are lightning and reflection. Lightning from a particular region in the early hours indicates a mountainous area, especially in the tropics. In polar regions a sharply defined patch of brightness in a gray sky is a sign of areas of ice floe or shore ice in the midst of open water.

b. *Indications by Sound*. Sounds from land may originate from the continued cries of sea birds from a particular direction, ships or buoys, and other noises of civilization.

c. *Other Indications of Land*. An increase in the number of birds and insects indicates nearby land. Seaweed, usually found in shallow water, may indicate the nearness of land. Bay ice, usually smoother, flatter, and whiter than pack ice, indicates a nearby frozen inlet, especially if the pieces are close together. Land is also indicated by odors that may be carried by the wind for long distances. This fact is important when navigating in heavy fog or at night. An increase in floating driftwood or vegetation means nearby land.

Figure 204. Bird noose.[18]

86. Taking Care of a Rubber Raft

(fig. 205)

a. General. Your chances of being a survivor of a downed aircraft are perhaps as likely as your being a survivor of a sunken ship. Aircraft carry rubber rafts. You should know how to care for it.

b. Proper Inflation. Be sure that your raft is properly inflated. If the main buoyancy chambers are not firm, use your pump or mouth inflation tube. Inflate cross seats if they are provided, unless there are injured men who must lie flat. Don't overinflate. Make air chambers well rounded but not drum tight. Remember that hot air expands. On hot days release some air.

c. Sea Anchor. Use your sea anchor (fig. 206), or improvise a drag from the raft case or bailing bucket to help you **maintain** direction and location, especially if you wish to stay close to the

---

[18]Aviation Training, Office of the Chief of Naval Operations, US Navy, "How to Survive on Land and Sea," Copyright 1943, 1951 by The United States Naval Institute.

6-MAN LIFERAFT
WITH SUNSHIELD

20-MAN LIFERAFT

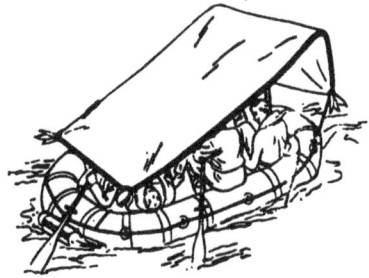

1-MAN LIFERAFT

Figure 205. Air Force rubber rafts.

wreckage of the ship or aircraft. Do not allow the anchor rope to chafe the sides of your raft. During a storm a sea anchor will help you to stay headed into the wind.

d. Snags. Be careful not to snag your raft. In good weather take off your shoes and tie them to the raft. Don't let fish hooks, knives, tin cans, and other sharp items cut your raft. Keep such items off the bottom of the craft.

e. Spray and Windshield. Keep your raft as dry as possible. Rig a spray and windshield in stormy weather. To keep the raft balanced, put weight in the center. If there are two or more persons aboard, let the heaviest sit in the middle.

f. Leaks. Leaks are more apt to occur at valves, seams, and on underwater surfaces. They can be repaired with repair plugs provided with the raft.

g. Sails. Never tie down both lower corners of a sail at the same time. A sudden gust of wind will tip your raft. Provide some method of releasing one corner of the sail. Hold it if necessary.

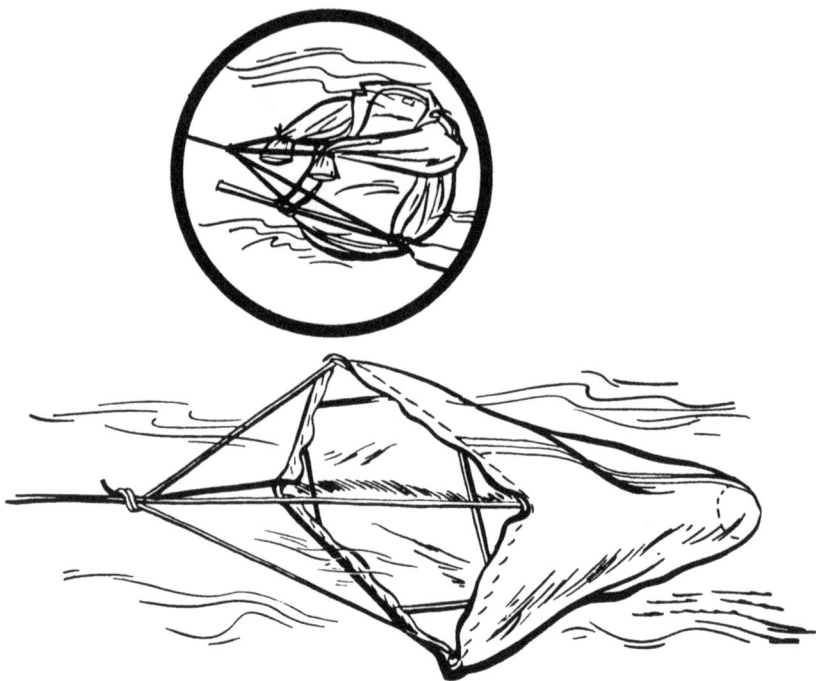

Figure 206. Sea anchor.

87. Signaling

a. General. You probably will have many ways to signal during the time you are surviving at sea—radios, flares, dye markers, mirrors, lights, and whistles. Don't use your signaling devices unless you are sure they will be seen or heard because a real chance of rescue may be missed. In the absence of signaling equipment, churn up the water with paddles or oars.

Caution: Make sure you are trying to get the attention of friends, not the enemy.

    (1) *Radio.* If your lifeboat or raft is radio equipped, follow the instructions for signaling and operating that come with it. Make sure you are out of enemy range before using your radio.

    (2) *Signaling mirrors.* See chapter 2.

    (3) *Lights and flares.* Instructions for the use of signal pistols, flares, smoke signals, and distress lights (normal lifeboat equipment) are found in watertight containers holding this equipment. The lantern is a valuable night light and can be used for signaling. The flashlight is also a good night signaling device.

    (4) *Signal flags.* The best method of displaying the signal

flag is for two men to stretch it taut by holding each end and moving it to present a flash of color. Signal flags flown from a mast can be seen from great distances.

(5) *Boat cover.* If you are using the tarpaulin or boat cover as a canopy, display it with the painted side up. Wave it when a rescue craft is in sight.

(6) *Whistle.* Use the whistle during periods of poor visibility to attract surface vessels or people ashore, or to locate other rafts when they become separated in the night.

b. *Avoiding Detection.* Take the following steps to avoid enemy detection:

(1) Travel during the night. Use your sea anchor during the day.

(2) Keep low in your raft.

(3) Stay covered with the blue side of the camouflage cloth up.

(4) Don't use your radio within 250 miles of enemy shores unless a friendly base is nearby.

(5) If detected and brought under fire by an enemy aircraft, be prepared to go overboard and submerge. If you are in a 20-man raft, go overboard and come up under the raft.

88. Seamanship

a. *Lookouts.* All men in a raft should serve on watch except the injured or sick. Arrange so that one man is on lookout duty at all times. Rotate this duty at intervals no longer than two hours. The lookout should watch for signs of land, friendly and enemy personnel, and signs of chafing or leaking of the raft. He should be lashed to the raft.

b. *Traveling.* Wind and current will drift your raft whether you want it to or not. Use this wind and tide if it is moving in the direction you wish to travel. To use the wind, inflate your raft fully and sit high. Take in your sea anchor and rig a sail. Use an oar as a rudder. If the wind is against you, put out your sea anchor and huddle low in the raft to offer less wind resistance. Don't try to sail your raft unless you know land is near. Don't worry about the currents. In the open sea they seldom move more than 6 to 8 miles a day.

c. *Raftsmanship.* Take every precaution to prevent your raft from turning over.

(1) In rough sea keep the sea anchor off the bow (front) and sit low. Don't stand up or make sudden movements.

(2) In extremely rough weather, keep a spare sea anchor ready in case you lose the first one.

(3) If your raft is capsized, toss the righting rope (on multiplace rafts) over the bottom. Move around to the other side; place one foot on the flotation tube and pull. If you have no righting rope, reach across and grab the lifeline on the far side. Slide back into the water, pulling the raft back and over. Most rafts have righting handles on the bottom. Twenty-man rafts need no righting since they are identical on both sides.

(4) To board a one-man raft, climb in from the narrow end, remaining as nearly horizontal as possible. This is also the proper way to board multiplace rafts when you are alone (fig. 207).

d. *Survival Swimming.* There is more chance of staying afloat if you relax, especially in salt water where the density of the water is more than the body. Trapping air in your clothing also helps. Always float on your back when possible. If you can't float, or if the sea is too rough, rest erect in the water. Inhale. Put your face down in the water and stroke with your arms. Then rest in this face-down position until you need to breathe. Raise your head, exhale, and support yourself by kicking your legs. Inhale and continue the cycle.

89. Health

(ch. 1)

a. *Mental Outlook.* High morale and reasonable resistance to some of the hazards of sea survival will bring you through. Don't let your imagination confuse your thinking. Relax and think out your problems. You have plenty of time. Take it easy.

b. *Physical Hazards.*

(1) A raft ailment that can become serious is immersion foot. This is caused by continued exposure to cool or cold sea water and poor circulation.

Figure 207. Righting a raft.

(2) Continued exposure to salt water may cause salt water burns or boils. Don't prick or squeeze these boils. Keep them dry.

(3) Sunburn and frostbite. See paragraph 61.

(4) Seasickness. Do not eat or drink if you are seasick. Try lying down and changing the position of your head.

(5) Sore eyes are caused by glare from the sky and water. Prevent this by wearing sunglasses or improvising an eye shield from a piece of cloth or bandage. If no medicines are available, moisten a piece of bandage, cotton, or cotton cloth with sea water and place this over your eyes before you bandage them.

(6) Constipation. Lack of bowel movement is normal on rafts. Don't be disturbed about it. Don't take laxatives even if they are available. Exercise as much as possible.

(7) Don't worry about the difficulty encountered in urinating or the dark color of urine. This is normal under such conditions.

(8) For a list of poisonous fish and sea life see chapter 7.

CHAPTER 7

HAZARDS TO SURVIVAL

90. Biological Hazards

a. Disease may be your worst enemy in your struggle for survival. Although it is not necessary for you to know a great deal about diseases, you should know about their presence in certain areas, how they are transmitted, and how to prevent them.

b. Most diseases are caused by minute parasitic plants and animals that enter the body, multiply, and set up a series of disturbances. Disease-causing organisms are placed inside your body by parasite transmitting agents. If you know which carriers are responsible for a particular disease, you are able to avoid contacting the disease by keeping the transmitting agents away from your body.

c. Some diseases usually occur only in certain areas. Therefore, supplement the general information in this chapter with information about biological hazards in your particular area.

91. Smaller Forms of Life

a. Don't Be Fooled. Many of the smaller forms of life can be more dangerous and uncomfortable than even a scarcity of food and water; however, their greatest danger is in their ability to transmit weakening and frequently fatal diseases through their bites.

    (1) Disease-transmitting agents require certain environmental conditions to exist and multiply, such as proper amount of sunlight, ideal temperatures, and suitable breeding sites. Because of these factors, you have only a limited number of disease transmitting agents to guard against at any one place or time. A transmitting agent that may be dangerous under certain circumstances in a particular area may be harmless under different circumstances in other areas.

    (2) Frequently, the particular disease organisms which are transmitted to man must, at some time in the course of

the transmitting agent's life, pass through one or more specific hosts. If the hosts are absent, the disease organism does not exist in the area and cannot be transmitted, regardless of how many potential transmitting agents are present. Man is a specific host in the case of malaria.

b. *Mosquitoes and Malaria.* Mosquito bites are not only unpleasant, but can lead to death. Whatever area you are in, you will probably encounter mosquitoes. They are more numerous in some areas of the Arctic and temperate regions in the late spring and early summer than they are at any time in the tropics. Tropical mosquitoes, however, are much more dangerous because they transmit malaria, yellow fever, dengue fever, and filariasis.

    (1) Malaria is transmitted to man only by the bite of some species of infected Anopheles mosquitoes. You can identify this mosquito, which is active only in the early evening or at night, because it rests with its tail end pointed upward at a 45° angle while biting. Its wings appear spotted (fig. 208).

    (2) Infested Anopheles mosquitoes bite an individual, injecting malarial parasites into the body of the victim. After a reasonable period of growth in the body, the parasites produce chills and fever by destruction of the red blood cells. At this stage the individual has large numbers of parasites in his bloodstream and another uninfested Anopheles mosquito takes a blood meal from the victim and subsequently injects the malarial parasites obtained in the blood meal into another victim.

    (3) Malaria may exist in any tropical climate where men live and where it is wet enough for mosquitoes to breed. It can be contracted during summer in may temperate regions but is not normally found in the cold climates of the northern and southern hemisphere.

    (4) The Aedes mosquito transmits yellow and dengue fever and bites at any time of the day or night (fig. 209).

        (a) Yellow fever is most common in West Africa, the Caribbean, and in parts of South and Central America. It can be prevented by immunization with yellow fever vaccine.

        (b) Dengue fever, transmitted like yellow fever, is weakening but seldom fatal. It is widespread in tropical and subtropical areas.

        (c) Mosquitoes also transmit filariasis, an infection which causes abnormal swelling of the body. This

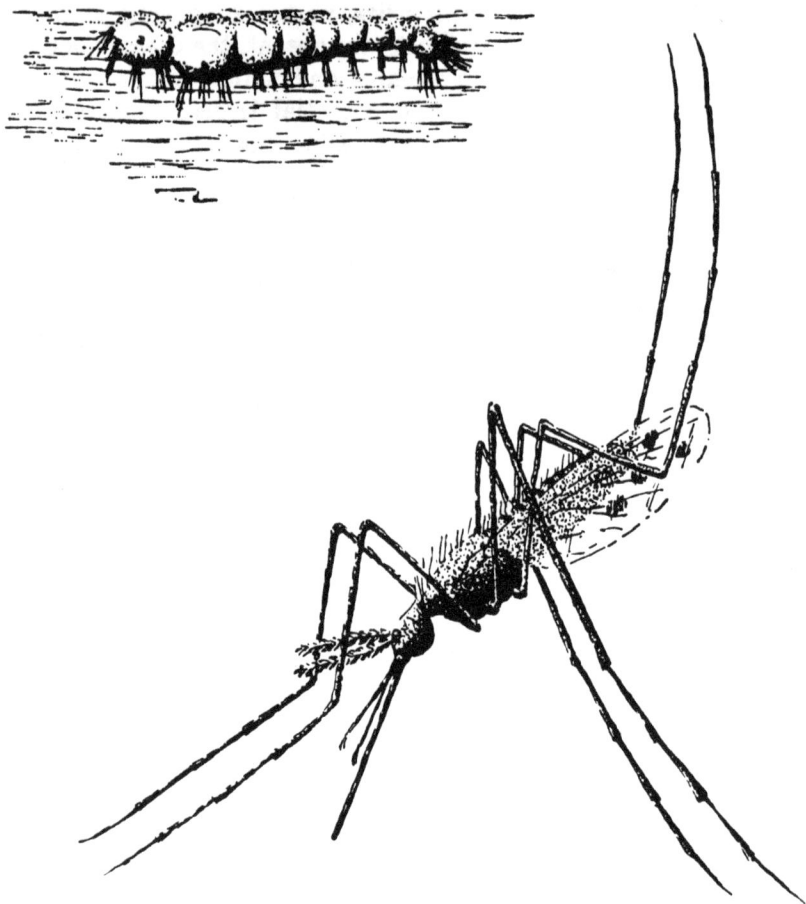

Figure 208. The Anopheles mosquito.

disease is found in both the tropics and subtropics.

(5) Take every precaution against mosquito bites. Follow these rules—

(a) Camp on high ground away from swamps.

(b) Sleep under mosquito netting. If none is available, coconut palm cloth, leaves, and other makeshifts will help.

(c) Smear mud on your face, especially before you go to bed.

(d) Wear all your clothing, especially at night.

(e) Tuck your pants into the tops of your socks or shoes.

(f) Wear a mosquito headnet and gloves.

(g) Use mosquito repellant, if available.

(h) Take antimalaria tablets according to directions as long as your supply lasts.

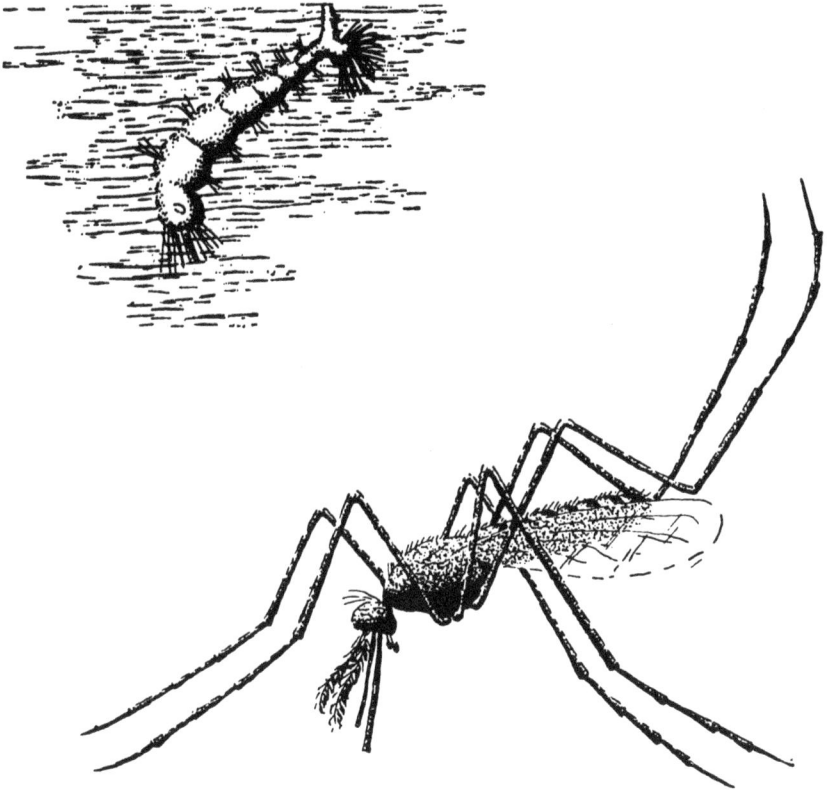

Figure 209. Aedes mosquito.

c. *Flies.* Like mosquitoes, flies vary in size, breeding habits, and in the discomfort or danger they can cause. The protection used against mosquitoes is generally effective against flies (fig. 210).

(1) Black flies are most bothersome in temperate regions but are found throughout the world. Some people react severely to the bites of these insects; more often, however, the danger is infection from scratching the insect's bite. Black flies may transmit filarial worms.

(2) Deer and horseflies are stout-bodied and usually light colored. They are seen during the day in areas where there are hoofed animals.

(3) No-see-ums or punkies are generally less bothersome than blackflies. They are tiny and inflict an itching bite. Some species may carry filarid worms. They are found throughout the world. If you find these gnats abundant, move on. They are often local and you seldom

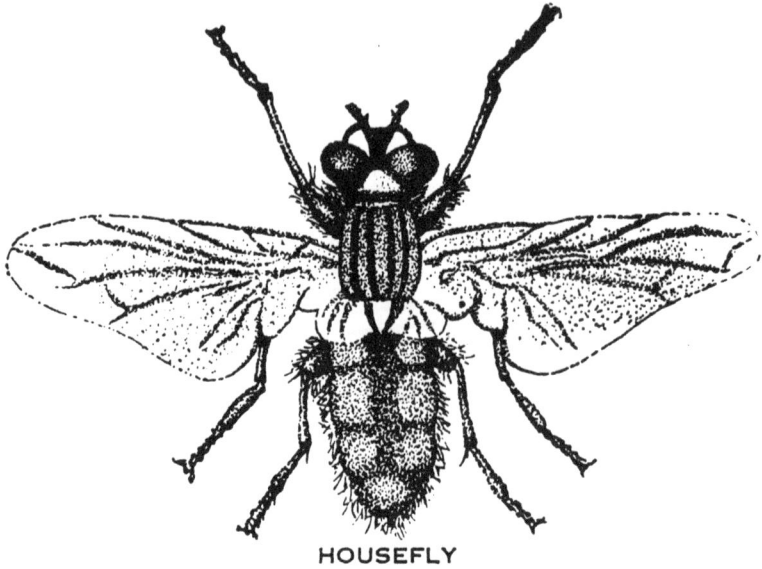

HOUSEFLY

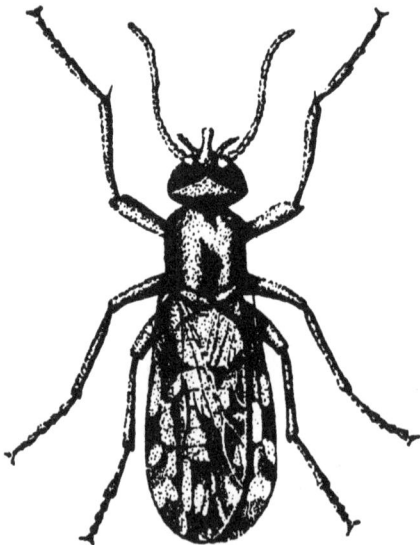

SANDFLY

TSETSEFLY

①

Figure 210.   Flies.

encounter them more than a half mile from their breeding areas.

(4) Sand flies are widely distributed, blood-sucking insects
and are suspected transmitters of various serious diseases. They can pass through ordinary netting, but

HUMAN BOTFLY

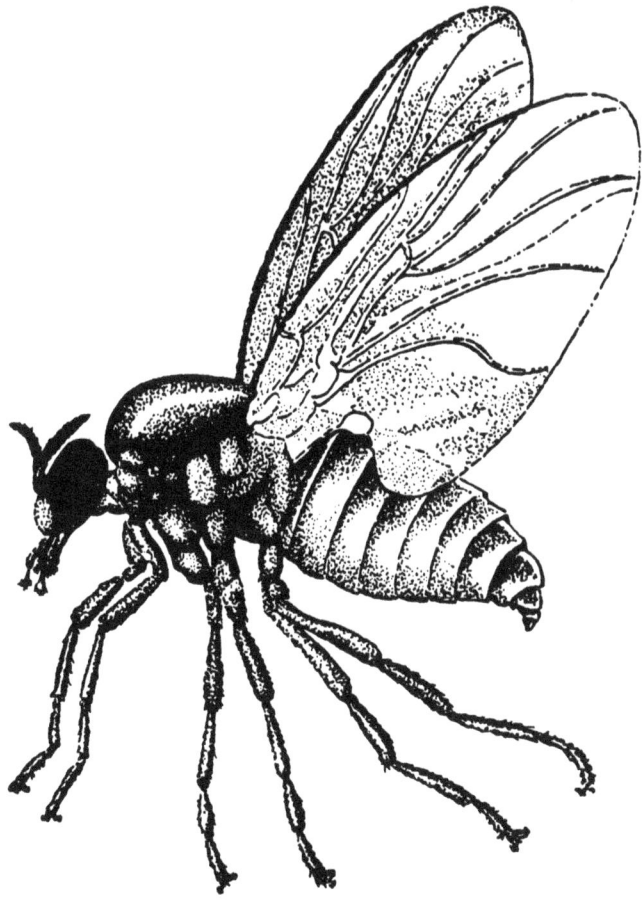

BLACKFLY

Figure 210—Continued

since they seldom fly more than ten feet above the ground and dislike air currents, they can be avoided.

(5) Eye gnats have a tendency to hover about the eyes and are potential carriers of dangerous eye infections. They also may transmit the serious syphilislike tropical disease called yaws.

(6) Tsetse flies are found only in central and south tropical Africa. Some species transmit sleeping sickness. All require shade and normally bite during the daytime. These insects prefer dark-skinned natives to whites.

(7) Screwworm flies are found in the Americas and southern Asia, especially the tropics. They are active during the day. Danger from the insect is greatest when you sleep in the open because the flies deposit their eggs in the nostrils, particularly if these passages are irritated by colds or wounds. The larvae burrow into the nasal tissues, causing severe pain and swelling.

(8) Blowflies are somewhat similar to the screwworm fly and are found in parts of Africa, India, Australia, and the East Indies.

(9) Bot flies are common in the Americas and the African tropics and are dangerous because of their larvae, which bore into the skin and cause a painful swelling that looks like a boil. Frequent applications of wet tobacco will kill the larvae, and you can squeeze them out.

  d. *Fleas.* These small wingless insects can be extremely dangerous in some areas because they can transmit plague to man after feeding on plague-ridden rodents. If you must use a rodent as food in suspected plague areas, hang up the animal as soon as it is killed and do not handle it until it becomes cold. Fleas will leave a cold body. To protect yourself against fleas, including the tiny chigoe, jigger, and sand flea, use repellent powder and wear tight fitting leggings or boots (fig. 211).

  e. *Ticks.* These flat oval pests are distributed throughout the world and are especially prevalent in the tropics and subtropics. They are carriers of relapsing fever and typhus. The two types of ticks are the hard or wood tick and soft tick (fig. 212).

(1) Never squash a tick on your skin or attempt to pull it out. Instead, cover it with a good coating of spit, alcohol, gasoline, kerosene, or iodine. The tick will free itself and be easy to remove. A lighted cigarette or match held close to the tick's body will also cause it to loosen its grip. If the tick's head remains embedded in your skin, remove it with the point of a knife blade which has been held over a flame.

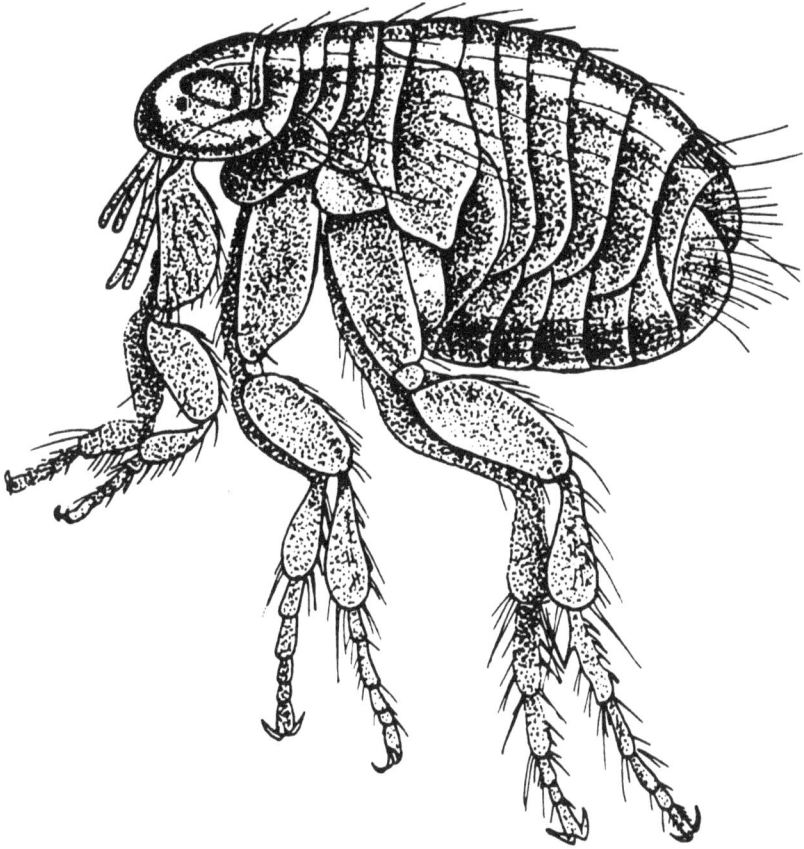

Figure 211. Flea.

(2) Rocky Mountain spotted fever, an acute infection which
may prove fatal, is tick-borne in the United States,
most of Canada, and in some parts of Mexico. This
disease is caused by a tiny living organism called a
rickettsia. Symptoms of the disease include a curious,
spotted rash, chills, fever, and severe pain, especially
in the arms and legs. Prevention is your best cure. In
suspected areas, wear protective clothing and inspect
yourself frequently for ticks.

(3) Diseases closely related to Rocky Mountain spotted
fever are tick-borne in the Mediterranean area, Brazil,
and elsewhere. Don't handle, prepare, or eat rodents
that are noticeably sluggish or sickly when killed.

f. Mites, Chiggers, and Lice. These very small insects are
common in many areas of the world and their ability to irritate
is entirely out of proportion to their size. Chiggers (fig. 213)

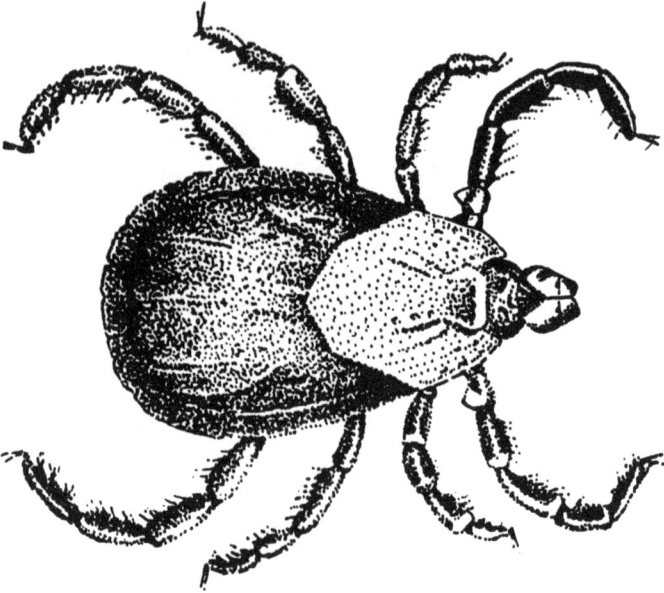

HARD TICK

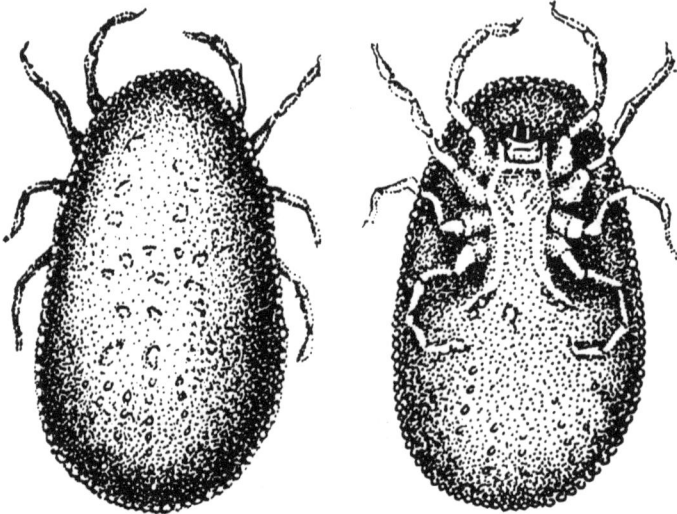

SOFT TICK

Figure 212. Ticks.

bore under the skin and cause itching and discomfort. People especially susceptible may be seriously affected and become ill. The human itch mite may cause various skin diseases such as scabies, Norwegian itch, and barber's itch. Native villages are usually infested with lice. Try to avoid huts and personal contact with the natives so far as possible. If you should be bitten by a louse, try not to scratch, because you will spread the louse feces into the bite. It is through infection with louse feces that man contracts such diseases as typhus and relapsing fever. If you do not have any louse powder, boiling your clothing will rid them of lice. If you cannot do this, exposing your clothes to direct sunlight for a few hours removes the lice. If you have been exposed to these insects, wash yourself with strong soap.

g. *Spiders.* Except for the black widow or hour glass spider of the southern United States, spiders in general are not particularly dangerous. Even the Tarantula is not known to bite with fatal or even serious effects. The black widow, however, along with tropical members of the same family, should be avoided as their bites cause severe pain and swelling and even death. All of these spiders are dark and marked with white, yellow, or red spots. Acute abdominal cramps may follow the bite of one of these spiders and may continue intermittently for a day or two. It is possible to mistake the pain for acute indigestion or even appendicitis (fig. 214).

h. *Scorpions.* The sting of this usually small animal is painful but seldom fatal. However, some of the larger species are more dangerous and their sting may result in death. Scorpions are found in widely separated areas and may constitute a real danger, since they hide during the day in your clothing, shoes, or bedding. Shake out your clothes before you put them on. If you are stung, use cold compresses or mud. In the tropics, apply coconut meat locally (fig. 215).

i. *Centipedes and Caterpillars.* Centipedes are numerous in the tropics, and some of the larger species can inflict painful bites. They seldom bite man, however, except when they cannot escape. They are not dangerous except when, like the scorpion, they have taken shelter in an article of clothing that is about to be worn. Caterpillars sometimes cause severe itching and inflammation if you brush against them (fig. 216).

j. *Bees, Wasps, and Hornets.* The stings of an aroused swarm of bees, wasps, or hornets may be dangerous, even fatal. Avoid nests as far as possible, but if you are attacked, plunge through some dense brush or undergrowth. The twigs springing back into position will beat off the insects (fig. 217).

(1) If stung, scrape off the barbed stings with a knife blade

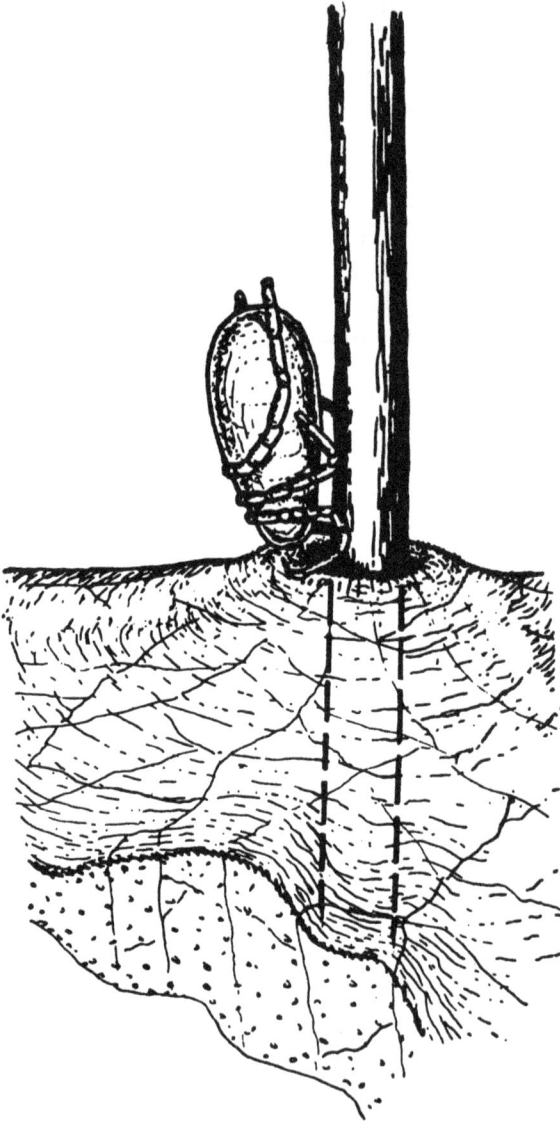

Figure 213.   Chigger.

to avoid forcing the poison into the wound. Ordinary mud helps to relieve the pain of a sting. The juice from the leaves of climbing hemp weed is a good antidote for stings. This is found near streams, swamps, and seashores in parts of America, Africa, and the South Pacific (fig. 218).

(2) Some tropical varieties of bees are large and militant

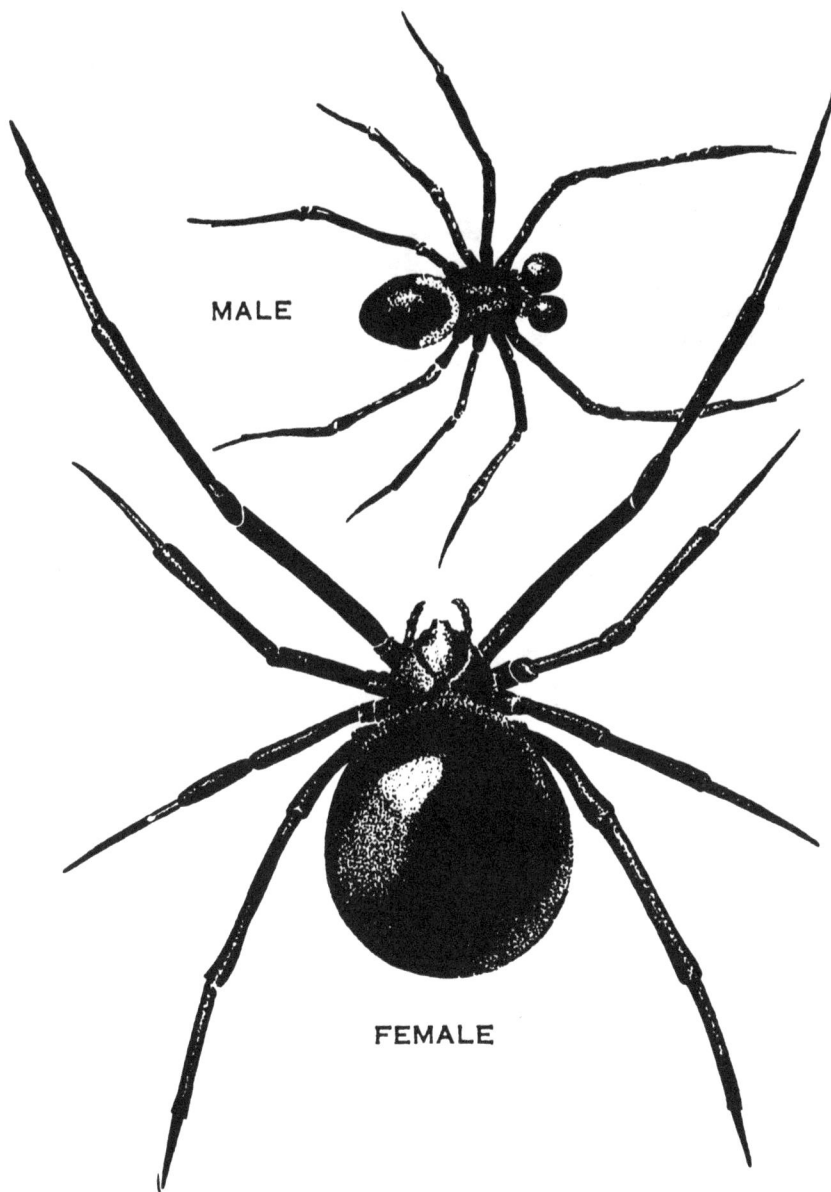

MALE

FEMALE

Figure 214. Black widow spider.

and are to be avoided. Use a smoke smudge to stupefy
the bees. Cover your head and hands, and you can take
the honey safely.

(3) Some tropical ants sting severely and attack in number.
Avoid them.

k. Leeches. These blood-sucking animals are found in such

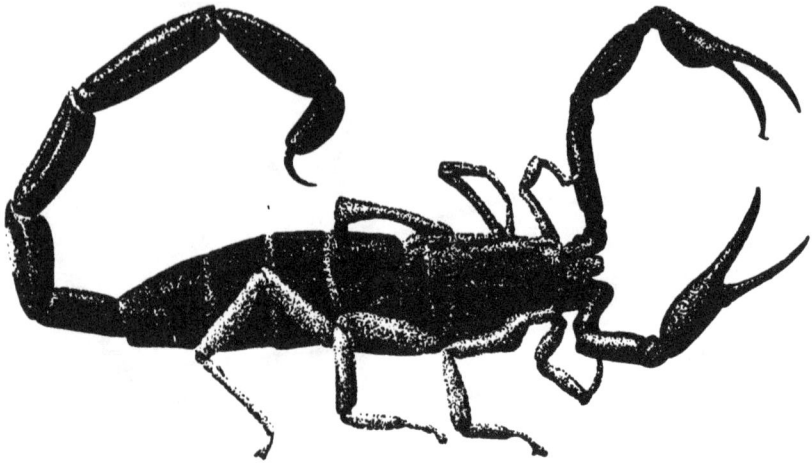

Figure 215. Scorpion.

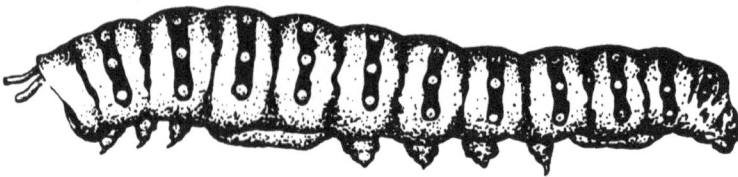

CATERPILLAR

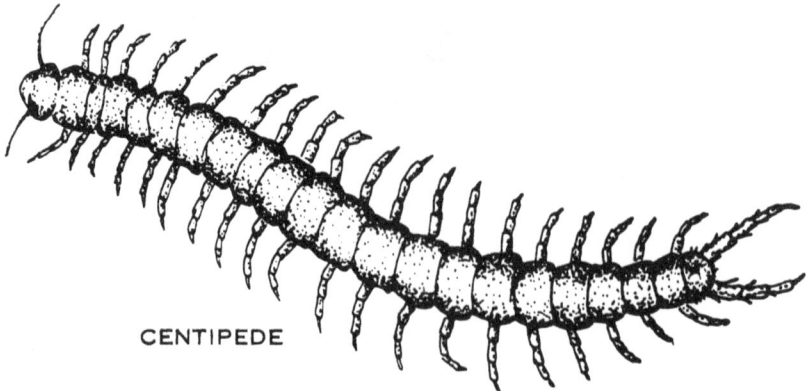

CENTIPEDE

Figure 216. Centipede and caterpillar.

widely separated areas as Borneo, the Philippines, Australia, the South Pacific, and various parts of South America. They cling to blades of grass, leaves, and twigs and fasten themselves to passing individuals. Bites cause discomfort, loss of blood, and may be followed by infection (fig. 219).

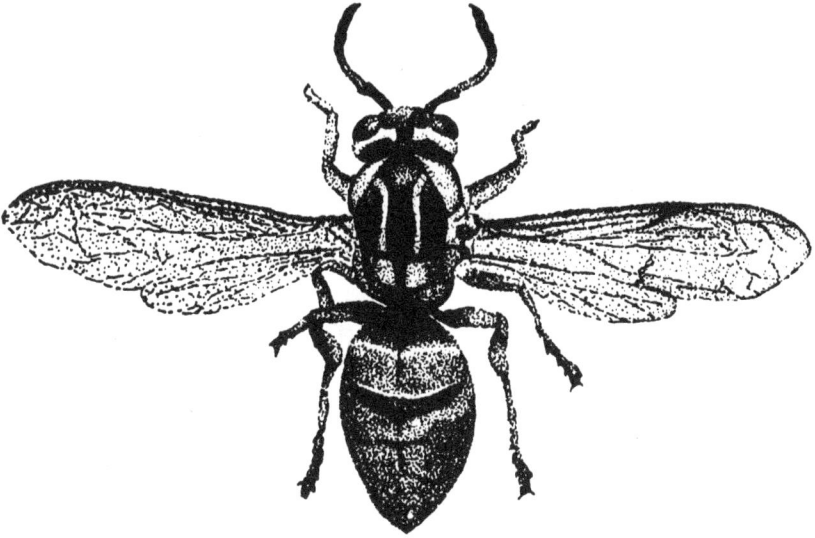

WASP

BEE

Figure 217. Bee and wasp.

(1) Remove leeches by touching them with a lighted cigarette, match, or moist tobacco. Place your trousers inside your boots and lace the boots tightly for protection.

(2) Leeches may be dangerous if they are swallowed in drinking water. Do not put your face into the water

Figure 218.   Climbing hemp weed.

if you drink from a spring—a leech may get into one of your nostrils. See chapter 3.

l. Flukes or Flatworms. These parasites are found in sluggish fresh water in parts of tropical America, Africa, Asia, Japan, Formosa, the Philippines, and other Pacific islands. There is no danger of flukes in salt water. Flukes penetrate the skin of those who come in contact with them either by drinking or

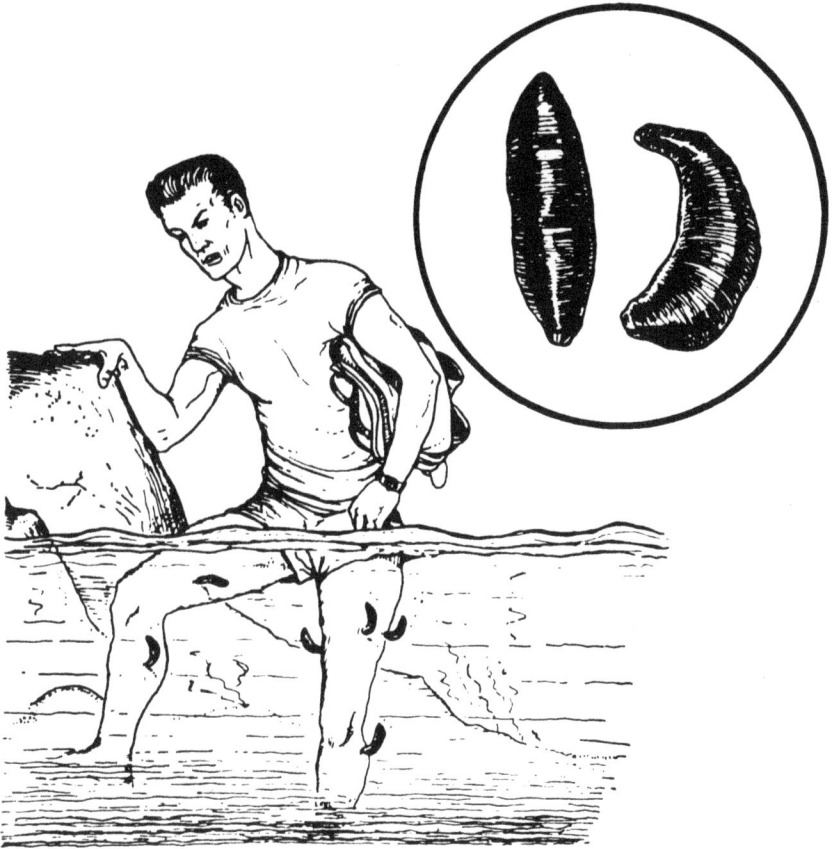

Figure 219. Leeches.

bathing in infested waters. They feed on blood cells and escape painfully through the bladder or feces (fig. 220).

m. Hookworm. Common in the tropics and subtropics, the hookworm larvae enter the body through the feet. There is no danger from hookworms in wilderness areas away from human habitation (fig. 220).

92. Poisonous Snakes and Lizards

a. Facts Outweigh Fiction. Fear of snakes is common among men, but comes more from improper training than experience. Actually only a small proportion of all snakes in the world are poisonous. While the variety of snakes is especially high in the tropics and decreases north and south of the Equator, danger from snakes in the tropics is less than in the rattlesnake- or moccasin-infested areas of the United States.

    (1) Generally, snakes get out of the way if given a chance, but you should follow these precautionary rules—

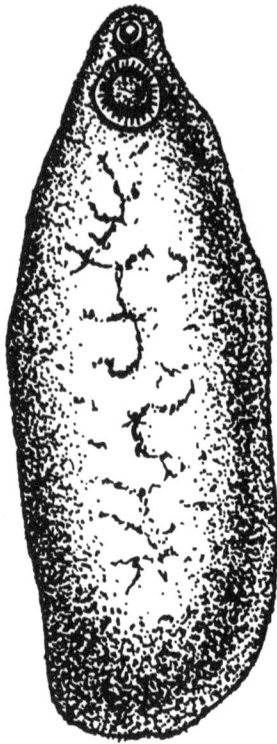

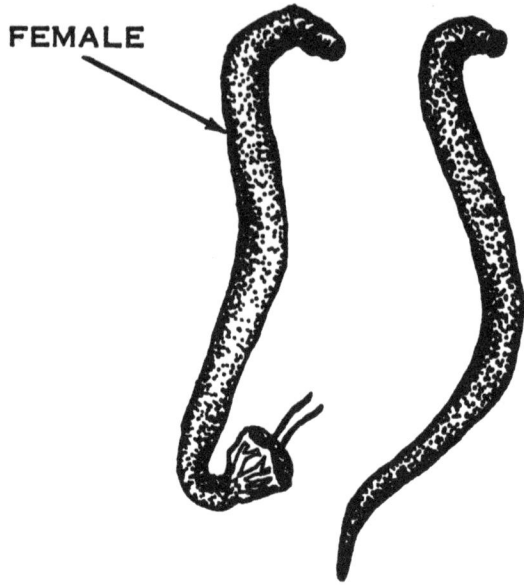

FEMALE

FLUKES   HOOKWORMS

Figure 220.   Flukes and hookworms.

(a) Walk carefully and watch where you step.

(b) When you are climbing rocky cliffs or picking up something from the ground, watch where you place your hands.

(c) Never tease or pick up a strange snake.

(d) Avoid sudden motion when you place your hand where a snake may be hiding.

(e) Wear leather boots, or at least loose clothing over your legs.

(f) Learn the habits and distribution of poisonous snakes in your area.

(g) Know what to do if you are bitten. See h below.

(2) Of the 2,400 different kinds of snakes, only 200 are dangerous to man. Snakes are found only in the temperate and tropical regions. Some areas of the world are free of poisonous land snakes, including New Zealand, Cuba, Jamaica, Haiti, Puerto Rico, and the Polynesian Islands.

b. Be on Your Guard. Some snakes are more aggressive than others and may attack man without apparent provocation. But aggressiveness is the exception.

(1) Snakes cannot stand weather extremes. In temperate regions they are active day and night during the warmer months; they hibernate or become inactive in cold weather. In desert and semidesert regions snakes are most active during early morning and evening and seek out shade during the day. Many snakes are active only at night.

(2) Snakes are normally slow travelers, but can strike with astonishing rapidity. They cannot outrun a man, and only a few can leap clear of the ground.

c. Poisonous Long-Fanged Snakes. Among the group of very venomous snakes are the vipers and adders of Europe, Asia, and Africa; the rattlesnakes, copperheads, and cottonmouth moccasins of North America; the bushmaster, fer-de-lances, and several other species of tropical America.

(1) The true vipers and pit vipers are mostly thick-bodied with flattened heads. Well known species of the true vipers, found only in the old world, are Russell's viper of India; the Cape viper of southern Africa; and the puff adder of dry areas of Africa and Arabia; and the gaboon viper of tropical Africa.

(2) The bite of a snake belonging to this group is very painful and is followed by local swelling which increases as the venom spreads throughout the tissue (fig. 221).

d. Poisonous Short-Fanged Snakes. Because of the relatively

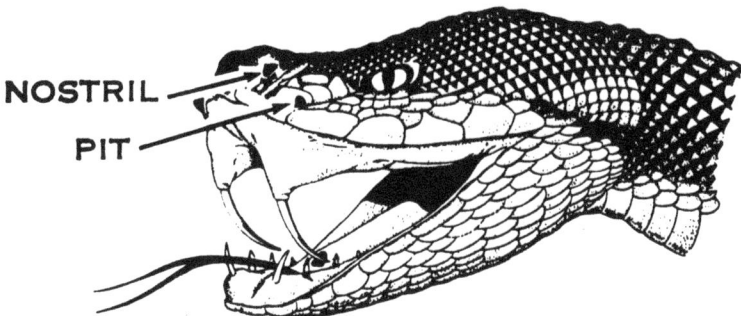

NOSTRIL

PIT

POSITIVE IDENTIFICATION FOR THE PIT VIPER FAMILY

Figure 221. Identifying a pit viper.

short fangs of snakes belonging to this group, even light clothing reduces their danger to man. This venom is the most deadly among poisonous snakes. Snakes included in this short-fanged group are the cobras, kraits, and American coral snakes. They comprise the majority of snakes in Australia, and many species are found in India, Malaya, Africa, and New Guinea (fig. 222).

(1) There are ten or more species of cobras, all found in the old world. All are more or less able to form a "hood." The king cobra is the largest of poisonous snakes.

(2) Many sea snakes belonging to this group occur in coastal waters of the Indian and Pacific Oceans. These snakes are not usually dangerous unless molested.

(3) The venom of the cobra and its relatives chiefly affects the nerves, and the cobra bite is not painful until some time later. Since the venom is promptly absorbed into the victim's bloodstream, it is distributed rapidly to all parts of the body.

e. *Sea Snakes.* Venomous sea snakes are not found in the Atlantic, but occur in large numbers off the shores of the Indian Ocean and the southern and western Pacific. They usually are encountered in tidal rivers and near the coast but may be seen far out at sea. They do not disturb swimmers, so there is little

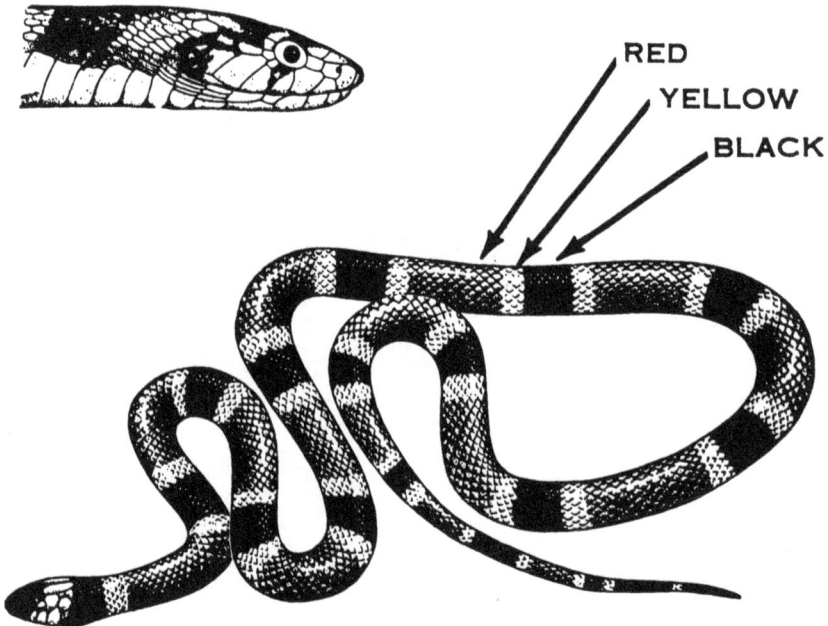

RED
YELLOW
BLACK

Figure 222. Coral snake.

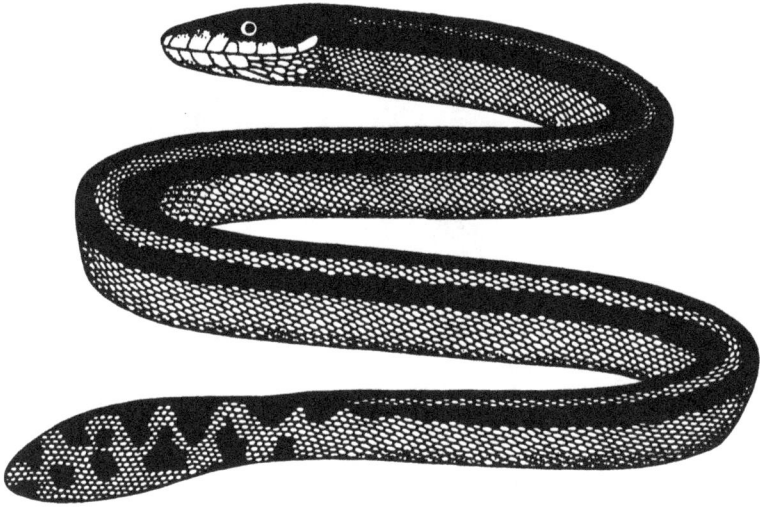

Figure 223. Sea snake.

danger of being bitten. They are identified by their flat, vertically compressed paddle tail (fig. 223).

f. Boas and Pythons. Boas and pythons are slow moving and timid and rarely attack man unless molested. Then they are vicious and dangerous because of their sharp teeth and power of constriction. The large species live only in dense jungle areas of the Philippines, southern India, China, South America, central and south Africa, Malay, Indo-China, and the areas of Burma (fig. 224).

g. Lizards. No lizards found anywhere in the world are poisonous except the gila monster and the beaded lizard, which *are found only in the American southwest (gila monster) and Central America and Mexico.* Because of the sluggishness of these lizards, they constitute little danger. Both are found only in desert areas (fig. 225).

h. First Aid for Snake Bite. Take no chance and treat all snake bites as poisonous. Follow these steps—

    (1) Place a constricting band between the bite and the heart. Apply constricting band just tight enough to produce distention of surface veins. Leave in place for one hour, then release for about one minute; reapply for about five or ten minutes; then release for another minute. Gradually increase the time "off" and decrease the time "on." This allows small amounts of poison to flow through the veins and increases its chances of being absorbed by the body.

    (2) As rapidly as possible make cross cuts across the bite

Figure 224. Boa.

Figure 225. Gila monster.

about ¼-inch deep and ½-inch long, sufficient to create a good blood flow. If you have no blade handy any sharp object will do (fig. 226).

(3) Suck or squeeze out the blood and venom.
(4) Remain quiet and do not move the bitten part.

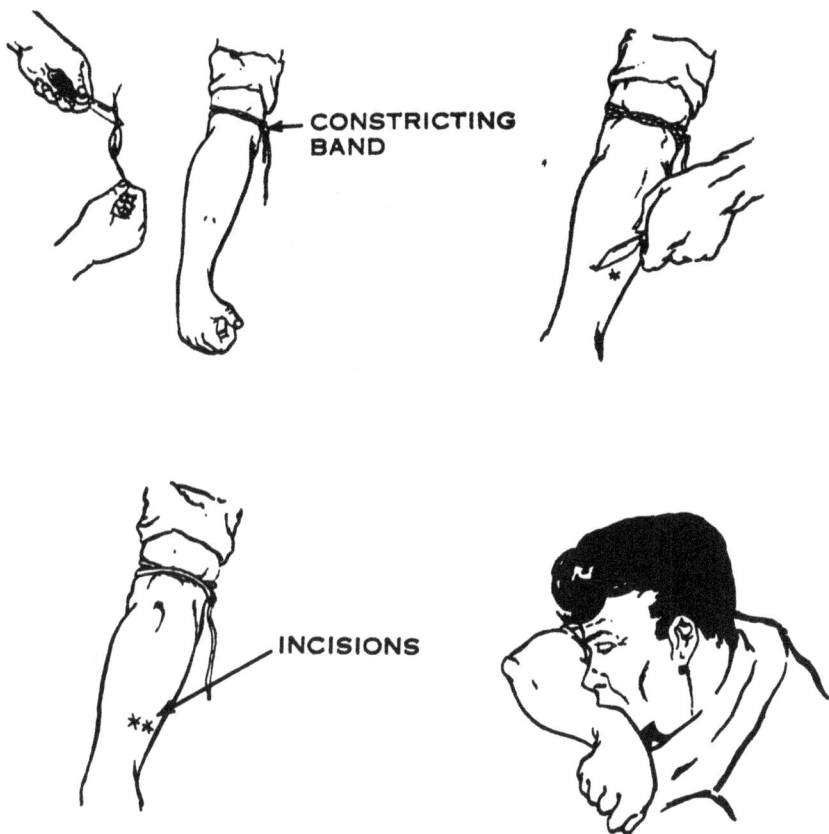

Figure 226. Snake bite incision.

(5) If ice is available place a cold pack around the bitten part.
(6) Act swiftly but remain calm.
(7) If after 15 minutes you feel no intense dryness and tightness of the mouth, headaches, pain or swelling of the bitten area, the bite is nonpoisonous.

93. Poisonous and Dangerous Water Animals

a. Sharks. These large aquatic predators are curious and will investigate objects in the water. It is unlikely they will attack unprovoked, but are likely to attack a wounded and bleeding swimmer. Any flow of blood should be stanched as quickly as possible. If you must go into shark-infested water, swim as quietly as possible (fig. 227).

b. Barracudas. The barracuda is found in most tropical and subtropical seas along reefs in murky water. It is considered

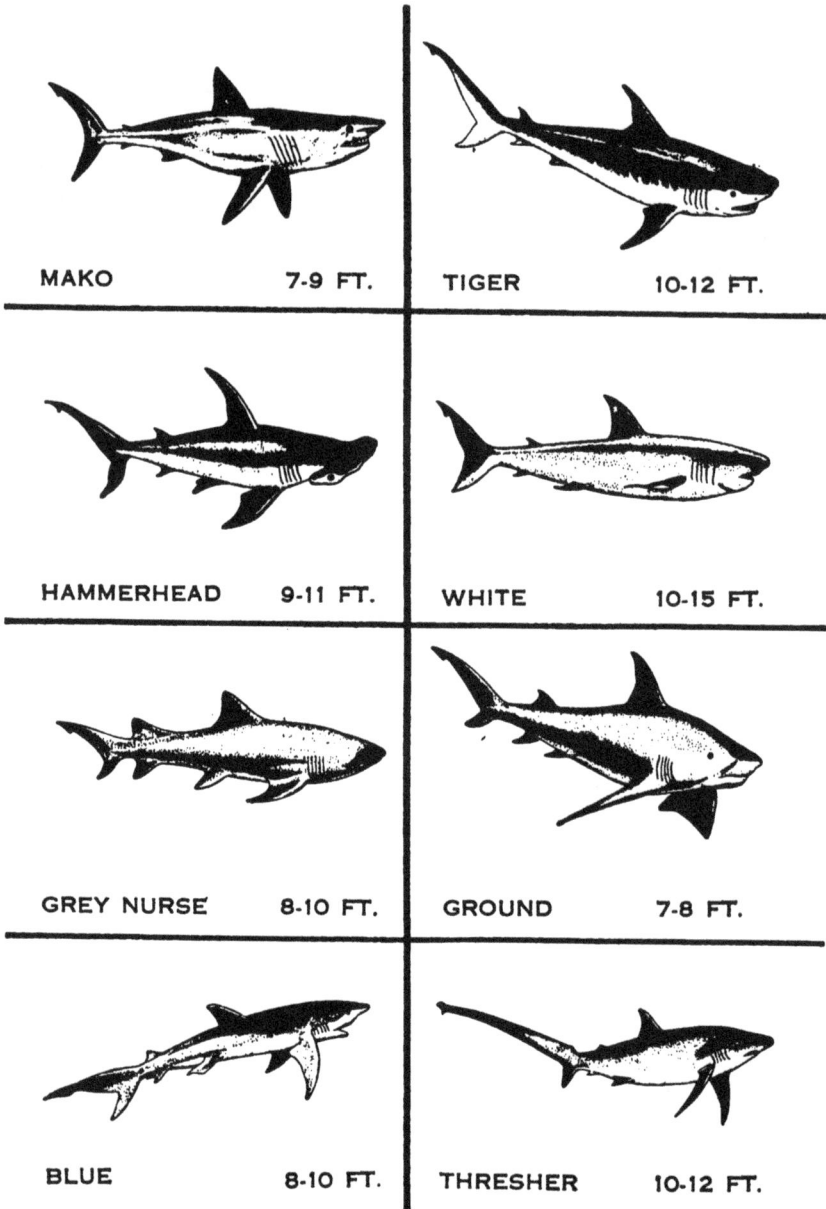

MAKO 7-9 FT.	TIGER 10-12 FT.
HAMMERHEAD 9-11 FT.	WHITE 10-15 FT.
GREY NURSE 8-10 FT.	GROUND 7-8 FT.
BLUE 8-10 FT.	THRESHER 10-12 FT.

Figure 227. Sharks.

by some to be more dangerous than a shark (fig. 228). The barracuda attacks indiscriminately.

c. *Electric Ray.* Found in both open water and along sandy and muddy bottoms, the electric ray, or torpedo, can give paralyzing shock. The torpedo, however, is rarely encountered (fig. 229).

d. Jelly Fish. The jelly fish, including the Portuguese man-of-war, is characterized by its ability to sting. The greatest danger is not from the sting itself, but from cramps that may develop in the swimmer. Clothing offers some protection from these animals, but if you are stung, try to relax. An application of ammonia relieves the pain of the sting (fig. 230).

e. Sting Rays. The sting ray has a poisonous spine on its tail that can cause a painful wound. These animals are flat, skatelike fish, often several feet in length. They are found in shallow, warm, coastal waters. Waders may clear a path in the water by poking a stick as they advance. *The sting of a large ray may be fatal.*

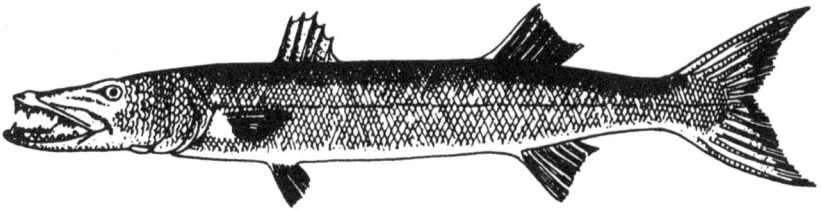

Figure 228. Barracuda.

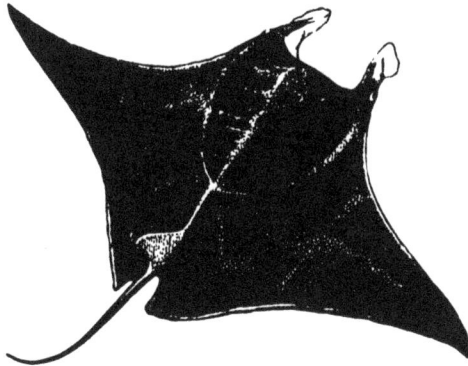

GIANT RAY OR MANTA

Figure 229. Ray.

Figure 230.   Jelly fish.

f. Scorpion, Toad, and Stone Fish. The stone and scorpion fishes of the Pacific Ocean and some of the toad fishes (fig. 231) of tropical America are the most dangerous poisonous fishes. They have stinging spines and may be encountered among coral. Treat a sting as you would a snake bite.

g. Other Water Hazards. Dangerous water animals listed

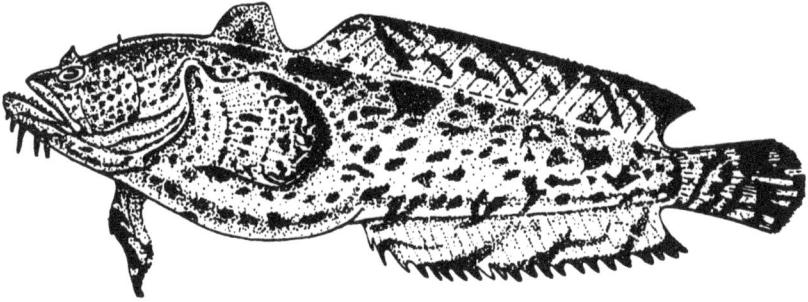

Figure 231. Toad fish.[19]

above by no means exhaust the list of hazards you may encounter. Tropical bone shell and long, slender, pointed terebra snails are also poisonous. Handle big conchs with caution; large abalones and clams can be dangerous if gathered by hand instead of pried loose with a bar or wedge. They may clamp onto your fingers and hold you under until you drown. Coral, dead or alive, can inflict painful cuts; seemingly harmless sponges and sea urchins can slip fine needles of lime or silica into your skin, which will break off and fester (fig. 232).

(1) Piranha are small fishes that live in the Amazon River and its tributaries in South America. Extremely strong, they are attracted by the least amount of blood in the water. An unprotected swimmer or wader may literally be reduced to a skeleton in a short time by a horde of piranha.

(2) Crocodiles and alligators are found in widely separated areas of the world. Alligators are found only in the southern United States and along the Yangtze River in China; crocodiles are found in coastal swamps, inlets and tidal rivers of the south Pacific, and in some areas of Africa and Madagascar. The American crocodile, found along coastal regions of Mexico, the West Indies, Central America, Colombia, and Venezuela, usually will avoid man. The crocodile is considered more vicious and treacherous than the alligator, but generally is not dangerous unless forced.

94. Danger From Mammals

Most stories about the dangers from larger animals are pure fiction. But very few animals will refuse to fight if forced into a corner. Many animals are dangerous when wounded or when

19Drawing from H. L. Todd, No. 22337, U. S. National Museum, collected at Pensacola, Fla. 1878, by Silas Stearns.

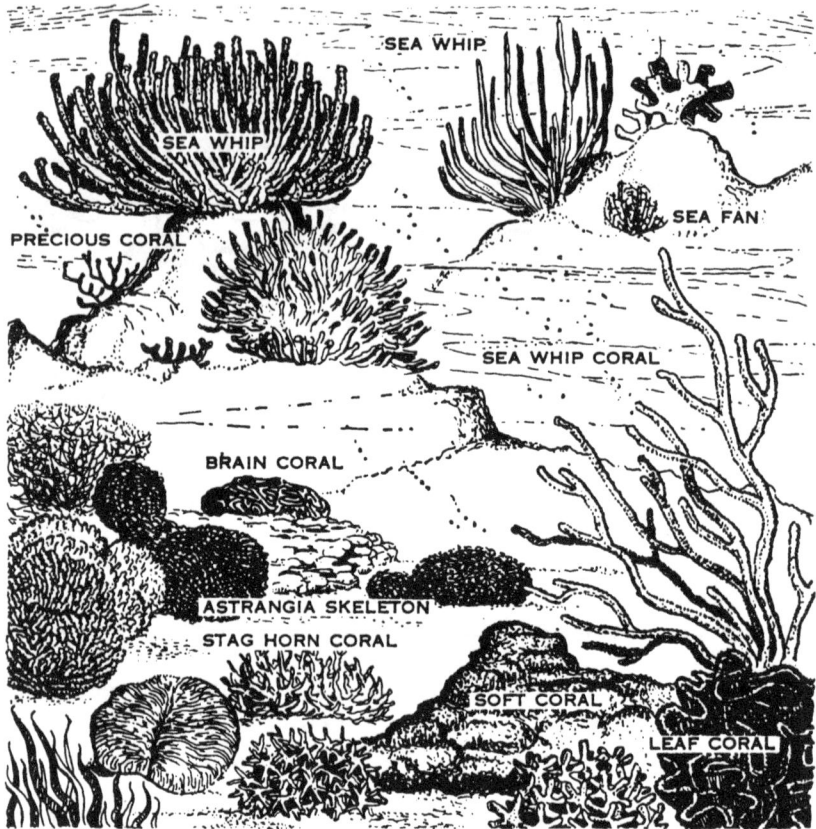

Figure 232. Shells and coral.

they are protecting their young. Old exiles or hermits such as elephants, boar, or buffalo that have been cast off by the herd are often cantankerous and belligerent. Lions, tigers, and leopards too old to hunt other animals successfully may become maneaters. Such animals are rare, however.

a. In Arctic and Subarctic regions bears are surly and dangerous. If you hunt them, don't shoot unless you are sure to kill. The polar bear rarely comes on land but is attracted by the smell of food caches or animal carcasses. It is a tireless, clever hunter and should be treated with great caution. The walrus too is a dangerous animal at close quarters.

b. Avoid wild buffalo because of their continued mean tempers. Approach wild pigs with caution. Elephants, tigers, and other large animals avoid man if given a chance, but they may charge when startled.

c. Bites from all canines (dogs, jackals, foxes) as well as some other meat eaters may cause rabies. Blood sucking vampire bats,

Figure 233. Vampire bat.

found only in South America, are not dangerous unless they are rabid or their bite becomes infected (fig. 233).

95. Poisonous Plants

a. Just Like Home. The danger from poisonous plants in other regions of the world is no worse than in parts of the United States. As a rule poisonous plants are not a serious hazard; but under certain conditions, they can be dangerous. The two general types of poisonous plants are those poisonous to touch and those poisonous to eat.

b. Plants Poisonous To Touch. Most of the plants poisonous to touch belong to either the sumac or the spurge families. The three most important poisonous plants in the United States are poison ivy, poison oak, and poison sumac. All of these plants have compounded leaves and small, round, grayish-green or white fruits. *If you learn the appearance and effects of these plants, it will help you in other parts of the world where similar plants flourish* (1 and 2, fig. 234).

    (1) Symptoms of plant poisoning are similar in all parts of the world—reddening, itching, swelling, and blisters. The best treatment after contact with these plants is a thorough wash using a strong soap.

    (2) There are many different kinds of plants in the tropics

POISON SUMAC

① Poison Sumac

Figure 234. Plants poisonous to touch.

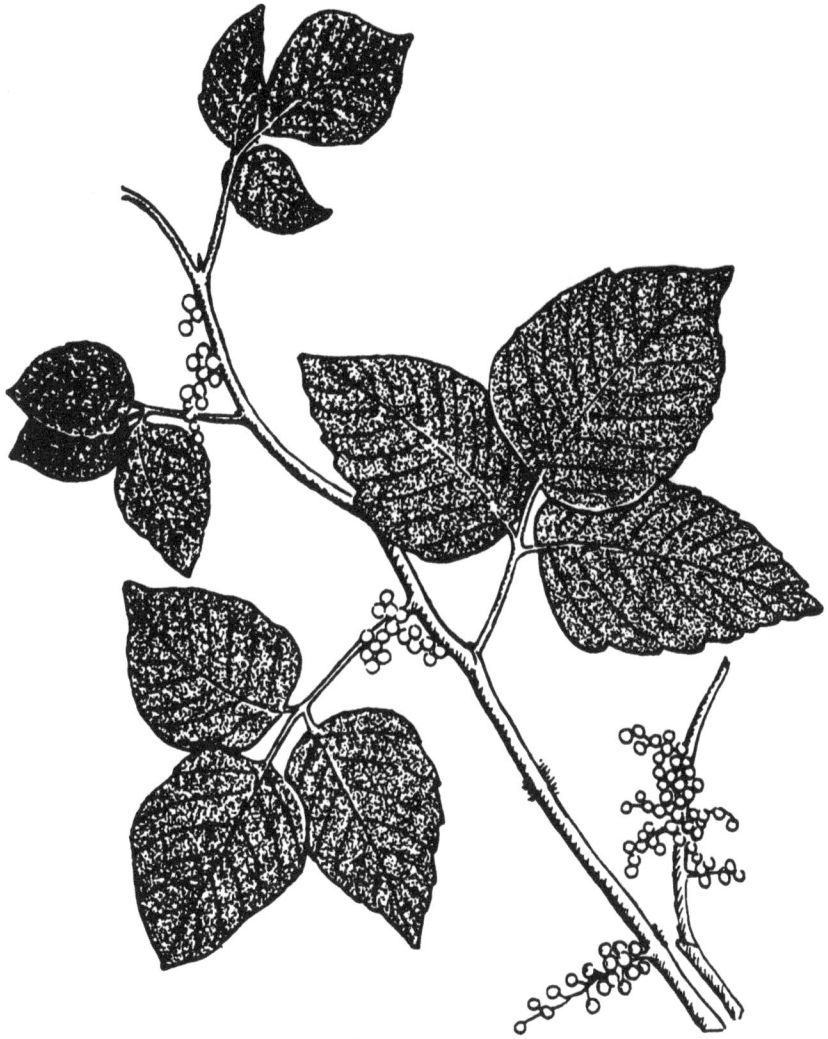

③ Poison Ivy

Figure 234—Continued.

and subtropics that are poisonous to touch. Plants more
frequently encountered are—

(a) The black poison wood of Central America.
(b) *Carrasco*, a shrub of the West Indies.
(c) The Rengas trees of Malaya, the Philippines, and
South Pacific Islands.
(d) The lacquer tree of China and Japan.
(e) Certain species of the Asian mangifera.
(f) The "blind eye," white mangrove found in Australia,
India, and the South Pacific Islands (fig. 188).

(g) The milky juices of such plants as the castor oil plant.

(3) When you are in a suspected poison plant area consider the following:

(a) Danger of becoming contaminated increases with overheating and sweating.

(b) Poisonous plant juices are particularly dangerous in the vicinity of your eyes.

(c) Milky juices of all unfamiliar trees should be avoided.

(d) Using the wood of any contact poisoning plant as firewood is dangerous.

c. Plants Poisonous To Eat. The number of poisonous plants is not great in comparison to the number that is nonpoisonous and edible. A good rule is to learn the plants that are edible (ch. 4) and, if necessary, eat strange plants in minute quantities and wait a while before continuing.

(1) In Arctic and Subarctic regions you may assume that less than a dozen plants are poisonous. Two of the most poisonous plants of the far north are water hemlock and poisonous mushrooms. See chapters 4 and 6.

(2) Poisonous plants are encountered in the tropics in no greater proportion than in the United States. If in doubt about which plants are poisonous and not poisonous, observe birds, rodents, monkeys, baboons, bears, and various other vegetable eating animals. Usually the foods these animals eat are safe for humans. Use these hints as a guide—

(a) Avoid eating plants that taste bitter.

(b) Cook all plant foods when in doubt. Except for the poison in some mushrooms (ch. 4), many plant poisons are removed by cooking.

(c) Avoid eating untested plants with milky juice, and do not let milky juice contact your skin. This rule does not apply to the numerous wild figs, breadfruit, papaya, and barrel cactus.

(d) Avoid ergot poisoning from infected heads of cereals or grasses by discarding grain heads having black spurs in place of normal seed grains.

d. Plants With Stinging Hairs. Plants with stinging hairs generally do not constitute a real danger. However, their sting, due to formic acid, is painful. Contact with stinging nettles found in waste lands of the United States and Europe will give you an example of what to expect from this type plant in other areas of the world.

APPENDIX I

REFERENCES

AR 320–50	Dictionary of United States Army Terms
SR 320–5–1	Authorized Abbreviations
DA Pam 108–1	Index of Army Motion Pictures, Filmstrips, Slides and Phono-Recordings
DA Pam 310-series	Military Publications Indexes
DA Pam 310–5	Index of Graphic Training Aids and Devices
FM 5–10	Routes of Communications
FM 5–20	Camouflage, Basic Principles
FM 21–11	First Aid For Soldiers
FM 21–22	Survival At Sea
FM 21–75	Combat Training of the Individual Soldier and Patrolling
FM 31–70	Basic Arctic Manual
FM 31–71	Operations in the Arctic
FM 31–72	Administration in the Arctic
FM 72–20	Jungle Operations
AF Manual 64–3	Survival
AF Manual 64–5	Survival
ADTIC Publication D–100	Afoot in the Desert
ADTIC Publication D–102	Sun, Land, and Survival
ADTIC Publication T–100	999 Survived

Aviation Training, Office of the Chief of Naval Operations, U. S. Navy, "How To Survive on Land and Sea," Copyright 1943, 1951 by The United States Naval Institute.

INDEX

[AG 353 (24 Jun 57)]

By Order of *Wilber M. Brucker*, Secretary of the Army:

MAXWELL D. TAYLOR,
General, United States Army,
Chief of Staff.

Official:
HERBERT M. JONES,
Major General, United States Army,
The Adjutant General.

Distribution:
   Active Army:

CNGB	AAA Bn
Technical Stf, DA	Cml Co
Admin & Technical Stf Bd	Engr Co
USCONARC	FA Btry
USARADCOM	Inf Co
OS Maj Comd	Ord Co
QM Tng Comd	QM Co
MDW	Sig Co
Armies	MP Co

Corps	Armor Co
Div	AAA Btry
Brig	Abn Co
77th USA SP Forces Gp	USMA
Bat Gp	Svc Colleges
Engr Gp	Br Svc Sch
Armor Gp	PMST Sr Div Units
Cml Bn	PMST Jr Div Units
Engr Bn	PMST Mil Sch Div Units
FA Bn	USA Arctic Indoc Cen
Ord Bn	USA Jungle Warfare TC
QM Bn	USA MT & Cold Wea TC
Sig Bn	Mil Mis
Armor Bn	ARMA
MP Bn	

NG: State AG; units—same as Active Army.

USAR: Same as Active Army.

For explanation of abbreviations used, see AR 320–50.

U. S. GOVERNMENT PRINTING OFFICE : 1959 O—491577

For sale by the Superintendent of Documents, U.S. Government Printing Office
Washington 25, D. C. - Price $1

E. W. SAWYER
425 North June Street
Los Angeles 4

FM 21-76
C 1

FIELD MANUAL

SURVIVAL

FM 21-76 } HEADQUARTERS,
DEPARTMENT OF THE ARMY
Changes No. 1} Washington 25, D. C., *19 October 1959*

FM 21-76, 25 October 1957, is changed as follows:

8. Aids to Maintaining Health
Protecting yourself against *** on your feet.

*　　*　　*　　*　　*　　*　　*

d. Guard Against Cold Injury.

*　　*　　*　　*　　*　　*　　*

(2) Frostbite is a *** the affected area. See **paragraph 61.**

*　　*　　*　　*　　*　　*　　*

9. Survival First Aid

*　　*　　*　　*　　*　　*　　*

d. Follow these procedures—

*　　*　　*　　*　　*　　*　　*

(2) *Bleeding.* Stop bleeding as *** the following methods:

*　　*　　*　　*　　*　　*　　*

(*d*) Apply a tourniquet *** lost blood volume. But remember—use a **tourniquet only when elevation or a pressure dressing over the wound fails to control bleeding, or when blood is spurting from a wound. (See paragraph 61*f*.)**

*　　*　　*　　*　　*　　*　　*

e. For treatment of cold injuries, see paragraph **61.**

14.1. Rate of Travel
(Added)

a. The rate of travel is based on the needs of your body, or the slowest man in the group, not on your desires. Plan and accomplish each day's travel so that you have enough time and energy left to establish a secure and satisfactory campsite.

b. Your rate of travel will be determined by a number of factors—
(1) The altitude of the area in which you are traveling.

(2) Your physical condition and that of your companion(s).

(3) The terrain (angle of slope and type of footing).

(4) The location of the enemy, his characteristics, number, and possible knowledge of your location.

(5) Time and distance requirements. Must you be at a certain place by a prescribed time?

(6) The amount of equipment carried. Carry only what is needed.

(7) Food requirements. If you hunt and gather food as you travel, there should be little need to undertake special travel to fulfill your need for food.

19. Mountains

*   *   *   *   *   *   *

e. When you are *** or even impossible. If necessary improvise **a rope** from **your** parachute **suspension** lines **and obtain a sturdy pole as a substitute for an ice ax.**

(1) *Rappelling* (fig. 11).

*   *   *   *   *   *   *

(*f*) Relax your grip *** off the rock.

*   *   *   *   *   *   *

2. Slow or stop yourself by tightening your grip on the ropes and bringing your **braking** hand across your chest.

*   *   *   *   *   *   *

(2) *Alternate method of rappelling.* If you have *** on a bight (fig. 12). **The man belaying should assume a secure sitting position with his feet braced.** The rope is passed low around **his** waist and is paid out as the **other** man moves. You must remember *** asks for it.

(3) (Added) *The hasty rappel.* Facing slightly sideways to the anchor, place the rope across your back, and under your arms. The hand nearest the anchor is your guiding hand and the lower hand does the braking. To stop, bring your braking hand across the front of your body, locking the rope. At the same time turn to face toward the anchor point. Use this rappel only on moderate pitches or on very long, gentle slopes. Its main advantage is that it is easier and faster to use, especially when the rope is wet.

f. (Superseded) As you travel down a grade, be on the lookout for slopes of loose, relatively fine rock. These slopes can aid your movement. It is best to descend them in a straight line. Here it

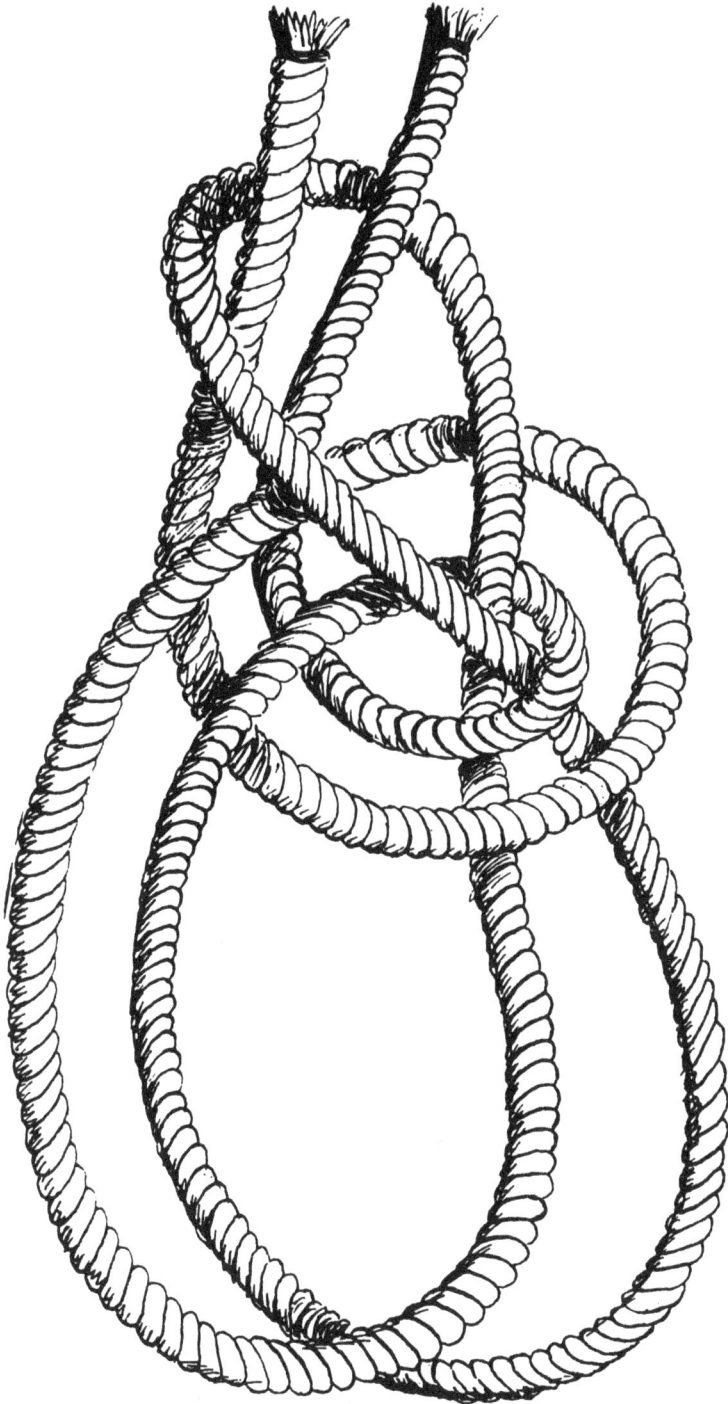

Figure 12. (Superseded) Bowline on a loop.

is important to keep the feet pointed straight down, the back straight, and the knees bent. Since there is a tendency to run, take care to avoid losing control.

20. Snowfields and Glacier Travel

*   *   *   *   *   *   *

b. If you are *** for the untrained (fig. 13). See **FM 31–72.**

*   *   *   *   *   *   *

21. Crossing Water

*   *   *   *   *   *   *

b. Methods of Crossing.
    (1) (Superseded) *Wading.* Before you enter the water, remove your socks and put your shoes back on. Do not risk the chances of having your feet cut by sharp rocks or sticks. Use a stout pole for support. Keep it upstream as much as possible as it helps to break the current. It also makes footing more secure and can be used to test for potholes.
    (2) *Swimming.*
      (*a*) Use the breast *** you begin swimming. If the water is too deep to wade, **lower yourself slowly to minimize the possibility of snags and falls due to obstacles hidden under the water.** In deep, swift *** with the current.

       *   *   *   *   *   *

    (3) *Rafts.*

       *   *   *   *   *   *

      (*b*) Spruce trees that *** 6 feet wide—

*   *   *   *   *   *

      6. (Added) Even with an ax, which is not likely in a survival situation, this type of woodwork requires a great deal of skill and time. A simple and more rapid method is the use of "pressure bars" lashed securely at each end to hold your logs together (fig. 20.1).

*   *   *   *   *   *   *

26. What Can You Drink?

*   *   *   *   *   *   *

b. (Added) For collecting rainwater, spread out clothing over sticks or limbs about 6 inches above the ground. Shape a sag in the middle of the cloth and put any kind of container under the lowest part of the sag. Most of the water will funnel to this sag

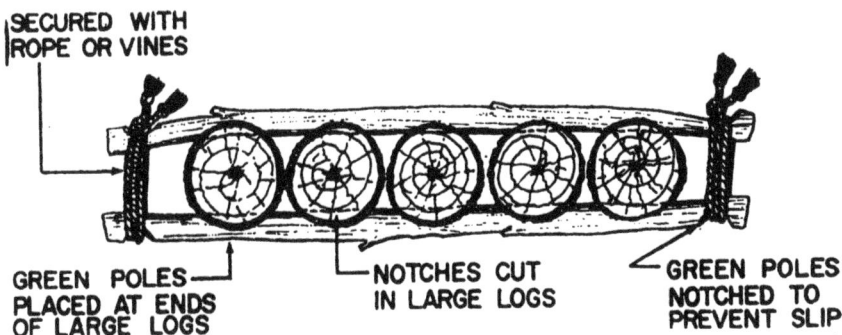

SECURED WITH
ROPE OR VINES

GREEN POLES
PLACED AT ENDS
OF LARGE LOGS

NOTCHES CUT
IN LARGE LOGS

GREEN POLES
NOTCHED TO
PREVENT SLIP

Figure 20.1 (Added) *Use of pressure bars.*

and collect in the container. The use of palm leaves or similar large leaves added along the edges of the clothing increases the collecting surface.

33. General

a. Experts estimate that *** a delicious meal.

Note. (Added) If in doubt as to whether or not a plant is edible, here are a few simple tests—

(1) If a plant contains a milky substance (sap), leave it alone.

(2) If you try to eat a plant raw and it has a burning, bitter, or nauseating taste, don't eat it.

(3) The *best* test is to boil the plant for 15 to 20 minutes. Place a small portion in your mouth and hold it there for 5 minutes. If there is no burning, bitter, or nauseating taste, swallow it and wait about eight hours. If there is no ill effect during this time, eat a handful. Wait another eight hours. If there still are no ill effects, the plant is safe to eat in any quantity.

*　　　*　　　*　　　*　　　*　　　*　　　*

34. Wild Plant Food

It is generally *** contain food value—

*　　　*　　　*　　　*　　　*　　　*　　　*

h. Fungi.

*　　　*　　　*　　　*　　　*　　　*　　　*

(3) Gilled fungi, or *** and poisonous varieties. **All fungi should be inspected. Only after firsthand knowledge, as in (d) below, should you consider them as valuable food sources. If there is any doubt, test the fungi as outlined in paragraph 33.** Supplement this information *** selecting edible mushrooms.

*　　　*　　　*　　　*　　　*　　　*　　　*

40. Birds and Mammals

*   ✿   ✿   *   *   *

c. Trapping.

*   *   ✿   ✤   ✿   *   ✿

    (2) *Trapping hints.* Following are some *** game or birds:

*   ✿   *   ✿   ✤   ✿   *

        (e) **Set snares or *** entrails for bait. After setting a trap in a runway, erect barriers on either side of it. These barriers should be made of dead branches, sticks and dry leaves. They are shaped to form a large "V" and funnel the animal into the trap. If the animal is moving slightly off the runway and should come in contact with the barrier, he will not jump over nor walk on it. Instead, he will travel parallel to the barrier and go straight to the trap. After erecting the barriers, you must do something about the human smell. Use animal blood or bladder contents to spread around the area. This will eliminate the human smell. When this is not possible, build a fire and smoke the area. The animal will not suspect anything after the area has been well smoked.**

*   *   *   ✿   *   *

        (9) (Added) *Bird snares.* The use of an effective snare for trapping birds that approach on both the wing and ground is to your advantage (fig. 131.1). The trigger is chisel shaped at the tree end and pointed at the perch end. The degrees of sensitivity may be adjusted by varying the depth of the trigger point in the bark. Birds that approach on the wing are snared by the feet. Those that walk to the snare will be strangled as they peck the bait.

46. Skinning and Cleaning

*   *   *   *   *

b. Fowl.

    (1) Most fowl should *** to pluck dry. **After the fowl is cooked, it can be skinned to improve flavor. Food value is lost, however.**

*   ✿   ✿   ✿   ✿   *

c. Animals.

    (1) *Skinning and dressing* (fig. 146). Clean and dress *** medium sized animals—

*   ✿   *   ✿   ✿   ✿

        (g) Save the kidneys *** and fleshy portions. **Check the**

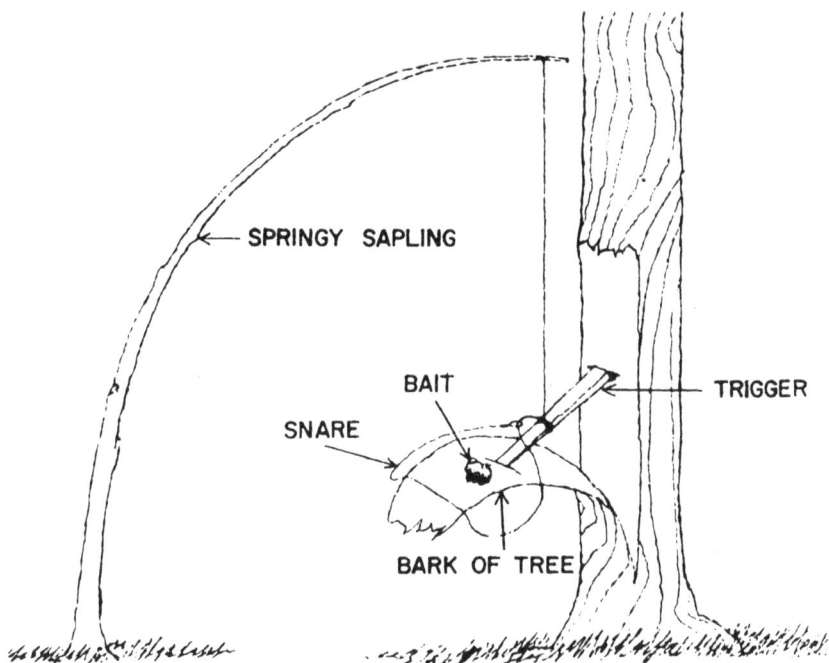

Figure 131.1 (Added) *Bird snare.*

heart, kidneys, liver, and intestines for spots or worms. If the animal is diseased in any manner, the greatest amount of danger is present while you are handling the meat and preparing it for cooking. If you have gloves, use them while preparing the animal. They will help prevent you from contracting the disease. Once the animal has been well cooked there is little chance of sickness, even though the animal was diseased.

(*h*) (Superseded) Do not throw away any part of the animal. The glands and entrails and reproduction regions can be used for baits in traps and fishlines.

*   *   *   *   *   *   *

(5) *Other edible animals.* All mammals are edible, regardless of what they are. Dogs, cats, hedge *** are especially valuable.

*   *   *   *   *

47. How to Cook

*   *   *   *   *   *   *

h. Cooking Plant Food. Soaking, parboiling, cooking or "leaching" are methods to improve taste. Of course, circumstances and the nature of the food dictate the method. Acorns can be made palatable by being crushed then "leached." (Leaching is done by crushing and pouring boiling water through the food while it is held in a strainer of some sort.)

*　　　*　　　*　　　*　　　*　　　*　　　*

(4) *Grains and seeds.* Grains and seeds *** be eaten raw. Grains and seeds may be ground into meal or flour.

*　　　*　　　*　　　*　　　*　　　*　　　*

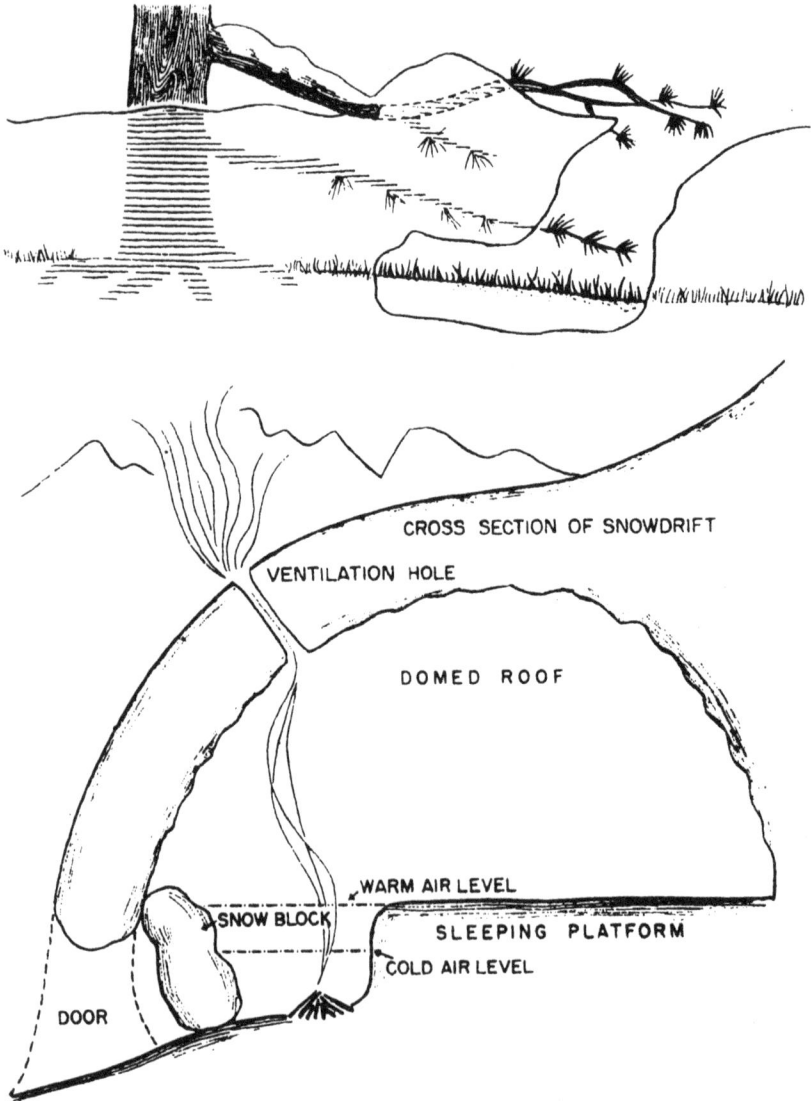

CROSS SECTION OF SNOWDRIFT

VENTILATION HOLE

DOMED ROOF

WARM AIR LEVEL

SNOW BLOCK

SLEEPING PLATFORM

COLD AIR LEVEL

DOOR

Figure 155. (Superseded) *Snow cave.*

8

i. Cooking Animal Food.

*   *   *   *   *   *   *

(3) *Fish.* Fish may be *** using a crane. **All fresh water fish and other fresh water foods should be thoroughly cooked, because they may contain disease-producing organisms.**

(4) *Reptiles and amphibians.* Frogs, small snakes *** stick, are edible. Skin all frogs and snakes before cooking. **The skin of a snake is not toxic, but its removal improves the taste of the meat.**

*   *   *   *   *

53. Climate and Weather

a. Temperature.

(1) *Arctic.* During the Arctic *** and frozen seas. Temperatures in the Arctic winter sometimes reach —70° F., **and range up to a maximum of 32° F.**

*   *   *   *   *   *   *

57. Water

*   *   *   *   *   *

c. It is safe *** observe these precautions—

(3) If you are hot, cold, or tired, **eating snow will tend to chill your body.**

*   *   *   *   *   *

58. Food

*   *   *   *   *   *   *

c. Fish. There are few *** may be plentiful. **Salmon are good to eat as long as they are alive when taken from the water.** However, after salmon spawn they are in poor condition and their flesh deteriorates. making them unfit to eat except as a last resort.

*   *   *   *   *   *   *

g. Plant Foods.

(1) Most plants in *** should be avoided (ch. 4). The water hemlock is one of the world's most poisonous plants. It can be distinguished by where it grows (always in wet ground) and by the following characteristics: a hollow, partitioned bulb at the base of the hollow stem, spindle shaped roots, and a strong disagreeable odor which is especially noticeable in the root and bulb. It is especially abundant in marshes near southern beaches and around marshy lakes in interior river valleys. It is never found on hillsides or dry ground.

(2) Some of the more common edible plants include—

*   *   *   *   *   *   *

(b) *Berries* (ch. 4). The salmonberry is *** are edible (fig. 172).

*   *   *   *   *   *   *

4. (Added) Wild rose (fig. 175.1). Fruit, called "hips," is available from midsummer through fall (often winter and early spring). The wild rose is found in dry woods, especially along streams and bluffs. It is distinguished by its prickly shrub. Hips are red to orange. In spring and winter, rose hips are hard and dry, but still edible and highly nutritious.

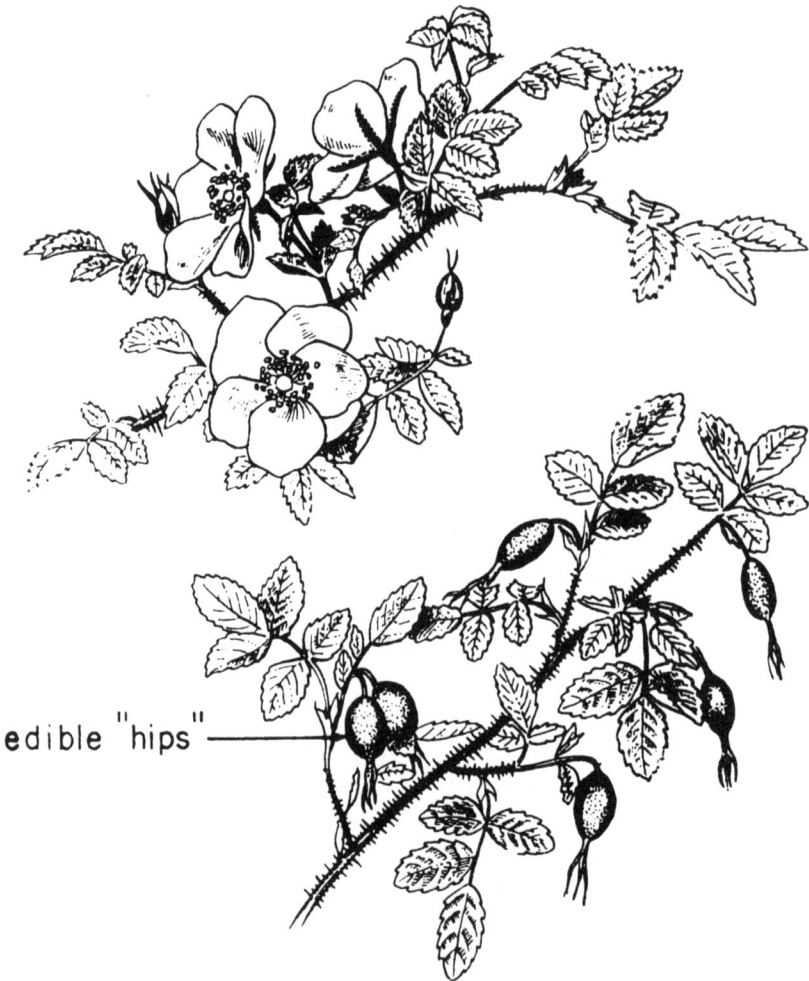

edible "hips"

Figure 175.1 (Added) *Wild rose.*

5. (Added) Other berries which can be eaten are the cloudberry and the crowberry (figs. 175.2 and 175.3).

(c) *Roots.* The following roots are edible:

          *   *   *   *   *   *   *

4. (Added) Licorice root (Eskimo potato) (fig. 178.1) has root-like tubers available in early spring, summer, and fall (occasionally winter). They become stringy and inedible in summer, however. The flowers of this plant are pink-purple, pea-like, and grow in elongated clusters; seed pods are flat, 1 to 2 inches long, and formed of several roundish joints.

(d) (Superseded) *Antiscurvy plants.* Scurvy can be prevented by eating fresh plants and meat. Many plants high in vitamin C content may be found, among which are scurvy grass (fig. 179) and spruce (fig. 180).

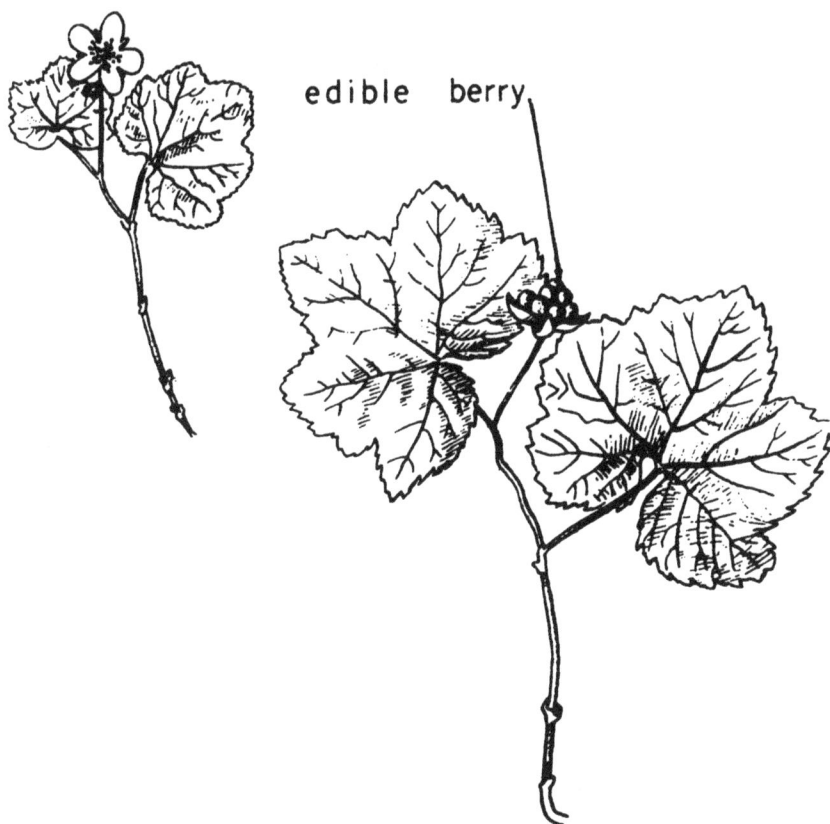

edible berry

Figure 175.2 (Added) *Cloudberry.*

(e) *Greens.* **Many northern** plants are good substitutes for the leafy vegetables you eat regularly as part of your normal everyday life.

*   *   *   *   *   *   *

5. (Added) Willow (fig. 182.1). These shrubs or small trees are found throughout Alaska. On the tundra, they may be only a few inches high. They have young, tender, leafy shoots that are edible during the spring. They become bitter and tough when old. Willows can be identified by their flower or fruit clusters that develop into a caterpillar-like spike an inch or so long. It is found in almost all habitats. It is also one of the richest sources of vitamin C.

6. (Added) Dwarf fireweed (Rock Rose). (See fig. 182.2.) The young leaves, stems, and flowers are edible in spring, becoming tough and bitter in summer. They are found along streams, sandbars, lake shores, and on Alpine and Arctic slopes. Stems are 1 to 2 feet tall, dying in the fall. Leaves are thickish, whitish, about

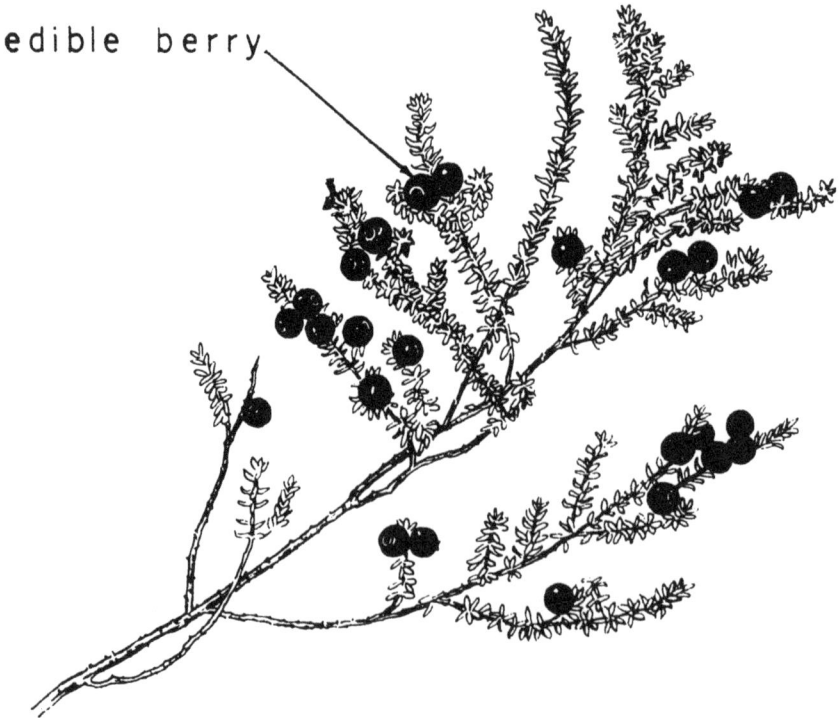

edible berry

Figure 175.3 (Added) *Crowberry.*

3 inches long. Flowers are rose to purple, large, showy, with four petals.

7. (Added) Tall fireweed (fig. 182.3). The young leaves, stems, and flowers are edible in spring, becoming tough and bitter in summer. This plant is found in open woods, on hillsides, stream banks, and near sea beaches. It is especially abundant in burned-over areas. It is similar to the dwarf fireweed, but its leaves are green and stems reddish and taller. It grows up to 6 feet tall; its flowers are showy pink.

Figure 178.1 (Added) *Licorice roots.*

Figure 182.1 (Added) *Willow.*

8. (Added) Coltsfoot (fig. 182.4). The leaves and flowering shoots are edible in spring and summer. The plant is found in moist woods and wet tundra. It has thickish leaves, triangular in outline and 3 to 10 inches long. dark green above and fuzzy white below. They rise from the ground only in the spring. The stalk is fleshy and cobwebby, about a foot high, with a cluster of creamy flowers at the top.

edible leaves,
   stems and flowers

Figure 182.2 (Added) Dwarf fireweed.

edible leaves,
stems and flowers

Figure 182.3 (Added) *Tall fireweed.*

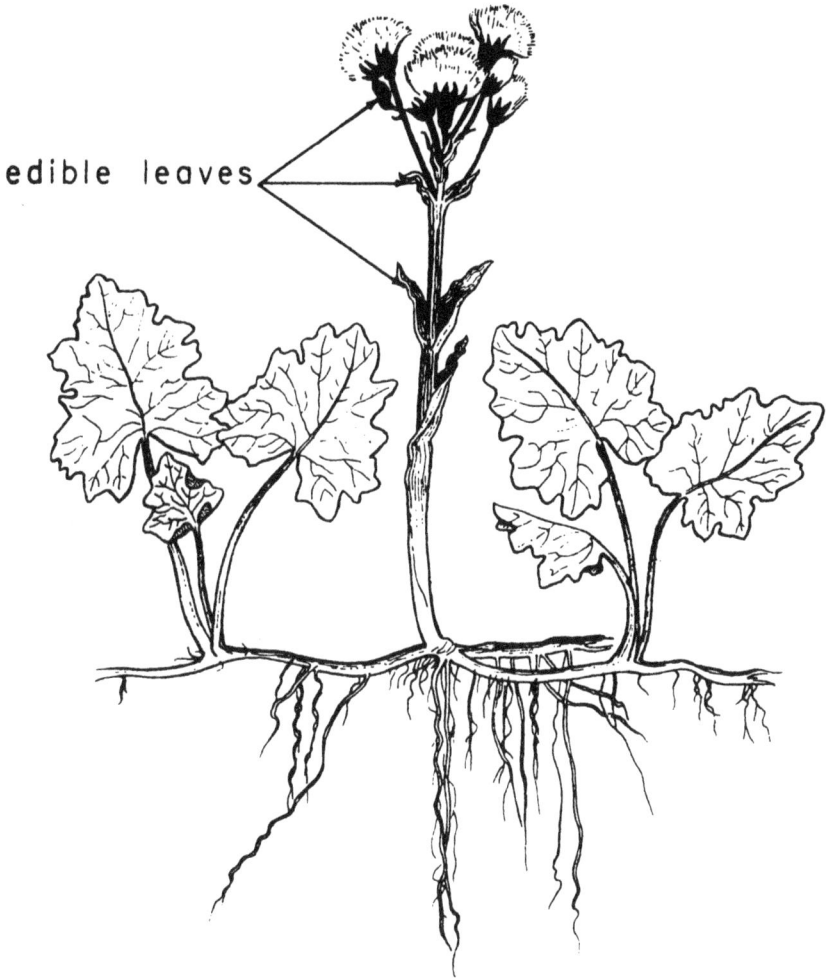

edible leaves

Figure 182.4 (Added) *Coltsfoot.*

60. Clothing

* * * * * * *

c. Some important facts *** to you are—

* * * * * * *

(5) You frequently will have to improvise some articles of clothing like boots, especially if your boots are too small to allow for extra socks. A piece of *** all you need. **The canvas seat of a military vehicle can be used when improvising boots, as well as can the seat pad cover of a pilot's parachute (fig. 184).**

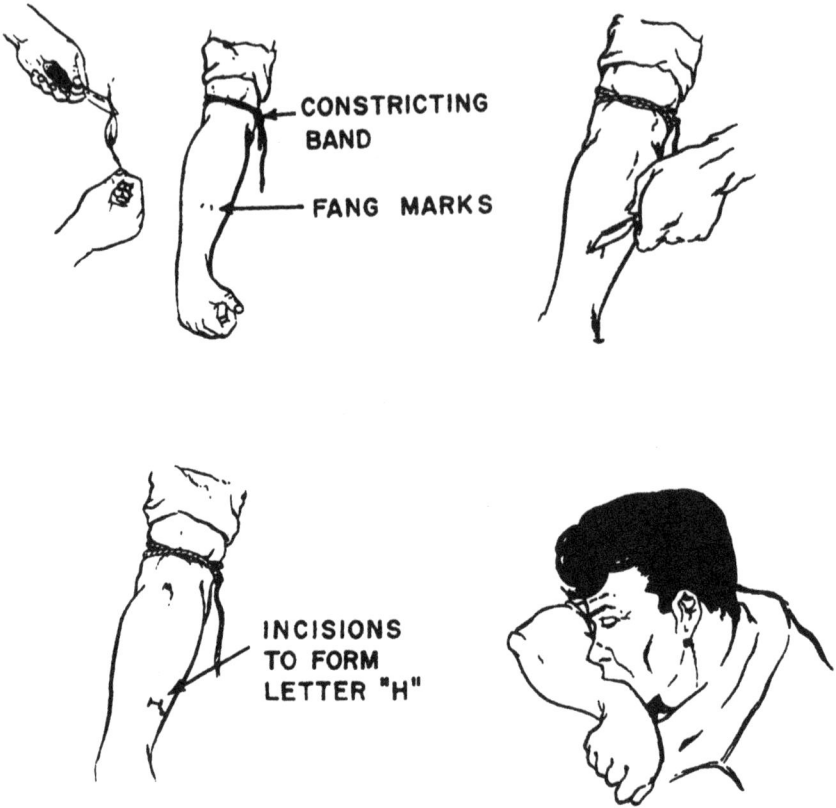

CONSTRICTING
BAND

FANG MARKS

INCISIONS
TO FORM
LETTER "H"

Figure 226. (Superseded) *Snake bite incision.*

61. Health

*   *   *   *   *   *   *

f. Use a tourniquet only to control severe bleeding from an arm or leg. **When elevation and a pressure dressing over the wound fail to control bleeding (or when blood is spurting from a wound!)** apply a tourniquet immediately. Once applied, the *** do not overheat.

*   *   *   *   *   *   *

83. Water

*   *   *   *   *   *   *

c. Observing water discipline.

*   *   *   *   *   *   *

(3) Do not drink urine. It is **unsatisfactory.**

*   *   *   *   *   *   *

92. Poisonous Snakes and Lizards

*　　　*　　　*　　　*　　　*　　　*　　　*

h. First Aid for Snake Bite. Take no chance *** follow these steps—

*　　　*　　　*　　　*　　　*　　　*　　　*

(2) As rapidly as possible make a cut across the bite in the shape of the letter H about ¼-inch deep and ½-inch long, sufficient to create a good blood flow. If you have *** object will do (fig. 226).

*　　　*　　　*　　　*　　　*　　　*　　　*

[AG 353 (24 Jun 57)]

By Order of *Wilber M. Brucker,* Secretary of the Army:

L. L. LEMNITZER,
General, United States Army,
Chief of Staff.

Official:

R. V. LEE,
Major General, United States Army,
The Adjutant General.

Distribution:

Active Army:

Tech Stf, DA (1) **except**	Engr Gp (3)
TQMG (15)	Inf Bg (5)
TAG Bd, US Army (2)	Armd Gp (5)
Chaplain Bd, USA (2)	QM Bn (3)
MP Bd (2)	Sig Bn (3)
Tech Stf Bd (2)	Armd Bn (3)
USCONARC (13)	Cml Bn (3)
USA Arty Bd (2)	Engr Bn (3)
USA Armor Bd (2)	Arty Bn (3)
USA Inf Bd (2)	ADA Bn (3)
USA AD Bd (2)	Ord Bn (3)
USA Abn & Elct Bd (2)	MP Bn (3)
USA Avn Bd (2)	Cml Co (2)
USA ATB (2)	Engr Co (2)
US ARADCOM (5)	ADA Btry (3)
US ARADCOM Rgn (5)	Arty Btry (3)
OS Maj Comd (10)	Inf Co (5)
OS Base Comd (5)	Ord Co (2)
Log Comd (5)	QM Co (2)
MDW (5)	Sig Co (2)
Armies (10) except	Armd Co (3) except
First US Army (12)	Armd Inf Co (5)
Div (5) except	MP Co (2)
Armd Div (14) (3 **ea cc**)	USATC (10)
Bde (3)	QM Tng Comd (10)